Joseph Wilson

Drainage for Health

Easy Lessons in Sanitary Science

Joseph Wilson

Drainage for Health
Easy Lessons in Sanitary Science

ISBN/EAN: 9783337390624

Printed in Europe, USA, Canada, Australia, Japan

Cover: Foto ©berggeist007 / pixelio.de

More available books at **www.hansebooks.com**

DRAINAGE FOR HEALTH.

Electrotypes of any of the Illustrations used in this book can be had at a moderate price upon application to the publisher.

A GENTLEMAN OF THE MALARIOUS COUNTRY VISITS THE CITY; AND AT THE SAME TIME GIVES HIS FAMILY A NICE CARRIAGE RIDE.—(*From a Photograph.*) See page 33.

DRAINAGE FOR HEALTH;

OR,

EASY LESSONS

IN

SANITARY SCIENCE.

BY

JOSEPH WILSON, M.D.

SECOND EDITION, WITH IMPORTANT ADDITIONS.

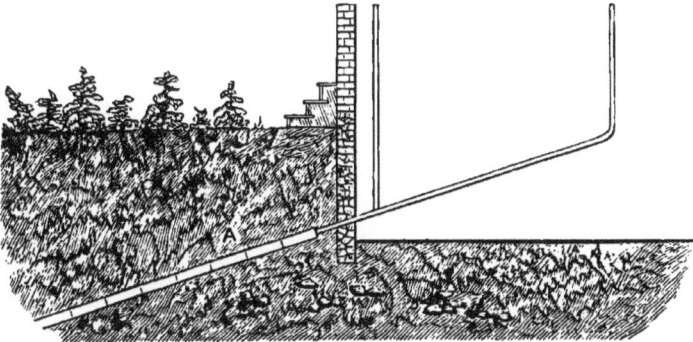

PREFACE.

THE present general interest in the subject of public health seems to call for such a work as is here attempted, simple in style and language, brief but correct as far as it goes, with sufficient variety to make easy reading. It is supposed that any gentleman may conveniently read it in leisure moments.

To the medical profession this summary may be found useful. Prepared with a view of meeting the wants of sanitary engineers and members of boards of health, farmers and legislators; if it should enable these to discuss its subjects intelligently, the object in preparing it will be satisfactorily accomplished.

PHILADELPHIA, 1886.

CONTENTS.

ILLUSTRATIONS.

ON DRAINAGE.

CHAPTER I.

THE DRAINAGE OF LAND.

(1) A YOUNG lady informs me that, being ambitious to possess a very elegant flower for her window, she planted a geranium in a small keg, an old paint keg, for want of a flower-pot large enough. This vessel neatly painted in white and brown stripes was gay enough. When the flower was first planted it looked pretty well, and she had hopes that it would excel everything of the kind in the neighborhood. But she was disappointed. Pretty soon there was a yellow leaf, next a dead leaf. The new growth was but a few very small leaves, while the old leaves wilted and died. Her plant was growing smaller every day, and it was made up principally of dead leaves and half-dead twigs.

The plant did not reach this condition without a few suggestions from friends. One thought that it was not watered enough, and this unlucky hint caused it to decline the faster. One was sure that it was not watered often enough; a little and often became the maxim, and still it declined. One imagined that it was watered too much, but this suggestion did not help the matter. The poor thing was nearly dead, just ready to be thrown on the dirt heap, when a more fortunate suggestion was made. A friend had a flower-pot with a hole in it, and when she was going to put a cork in the hole to prevent it from soiling the window, a gardener told her not to do so or she would kill the plant. After a long consultation a gimlet was borrowed, three holes were bored in the keg, and after this experiment, with so little of hope or apparent reason about it, the plant began to grow. There were new leaves, handsome flowers, larger branches. The paint keg with its load of flowers became the admiration of the neighborhood.

(2) The clay soil of a meadow, like the bottom of the paint keg, does not allow the water to penetrate or filter through. Partly the rainfall flows over the surface, washing it into gullies, and partly it evaporates; but the crops never flourish on this clay soil. The farmer, when he prepares his field for wheat, tries to remedy this by ploughing his land into ridges and hollows, thus making surface drains. The earth to the depth stirred by the plough is permeable to water, which thus finds its way to the drains without washing the surface so much into gullies. · The wheat grows best on the middle of the ridge, the highest part, and scarcely any grows in the hollows. This peculiarity of the growing crop is

FIG. 1.

Wheat on Heavy Land, as Commonly Cultivated.

seen all over the field. It is seen not only on marshy land, but on any ordinary soil resting on tolerably pure clay. Some plants can hardly be made to grow at all on such land; but with such drainage as we have just described—the ridges and surface drains—

FIG 2.

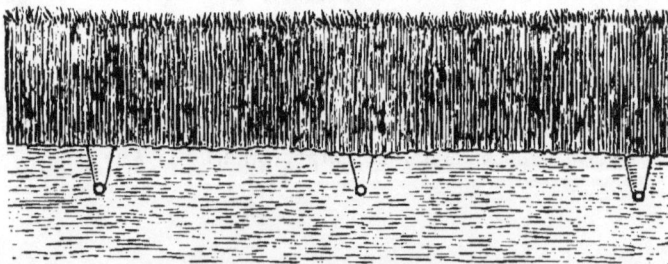

Wheat on the same Land Properly Drained.

wheat and grass crops, having their roots near the surface, do pretty well; but fruit trees and grapevines, having deeper roots, do not flourish without deeper and more effective drainage. Sometimes this is done by channels of earthen pipes, three or four feet beneath the surface.

(3) The physiologists have not told us why these holes in the flower-pot are so essential, or why this clay soil is so preventive of crops. It cannot be mere wetness or dryness, for that kind of difficulty would be obviated by a suitable supply of water. Virgil encourages us with the suggestion that it is a beneficent arrangement of Divine Providence to make us sharp, a sort of whetstone, *curis acuens mortalia corda.* Palladius and Columella, a century later, describe devices still in use for draining marshy fields. They had open ditches; they had covered drains made by filling the ditches half full of brush or stones, and covering them up with earth, as in what we call French drains. These devices must have been quite effective, for the Roman Campagna, where Cincinnatus followed his plough, is covered with the ruins of substantial houses, to say nothing of palaces. In modern times this

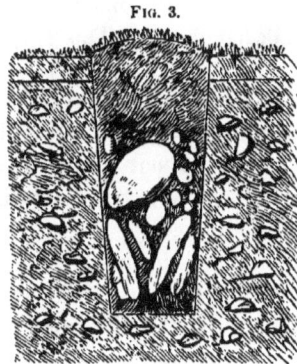

Fig. 3.

The French Drain of the Roman Plains.

Campagna is about the most pestilential spot on the face of the earth; the drains were neglected; the inhabitants died of *malaria;* the region is a desert, with a few huts among the ruined walls of ancient villas. Rasori studied the cause of this death and desolation; he called it *malaria.* Garibaldi engaged in the gigantic enterprise of restoring these ancient drains, of draining this marshy land, so as to render it healthy, fertile, habitable.

(4) But we need not go so far as Italy to find land thus desolate for want of drainage, nor need we seek for flat and marshy land. *Any piece of land with its substratum of impermeable clay, no matter how manured or how watered, is unproductive of vegetation and unfavorable to health until it is provided with effective drainage.* The moisture, much or little, stagnating in the soil poisons vegetation. It has been suggested that the plants secrete poisons into the soil,— a poisonous excrement,—which, unless removed by percolating water, poisons the plants themselves; and this is, perhaps, the only explanation that has been offered.

There certainly is a very deadly poison about such land, very destructive of human life. This poison we know only by the

havoc it makes, the disease and death which it causes. It is called
malaria—malarial miasm, malarial poison, marsh miasm, marsh
fever, poison, etc. The diseases that it principally causes are various types of *periodic fever*,—intermittent, remittent, congestive,
pernicious; and malarial anæmia and diarrhœa, ague-cakes and
dropsy, jaundice and marasmus. Excepting the deaths of infancy
and old age, probably one-third of all the deaths in the world are
caused by malarial poisoning. During the first two years of our
civil war 71,192 deaths were reported from all causes, and 20,675
of these deaths, more than two-sevenths ($\frac{2}{7}$) of them, were from
malarial fevers. This is, perhaps, a fair average in a protracted
war in a healthy country. Special service in less healthy countries
causes much greater mortality. In 1809 a British army of 39,214
men embarked for Holland, and was nearly destroyed in four
months. About nineteen-twentieths ($\frac{19}{20}$) of all the deaths and
disabilities were from malarial disease.—BLANE.

(5) By keeping away from malarious places we may often avoid
disease and death. We may sometimes do better by draining the
land and making it healthy, thus increasing the crops two or three
fold, and greatly lessening the labor of cultivation. The arrangements for drainage must vary according to the locality and character of the soil, thus:

Natural Subsoil Drainage.—In a hilly country with light soil
not much need generally be done. When the forest is first cleared
off and the soil turned up, the place is likely to be unhealthy; but
after the land is fairly cleared and ploughed into ridges and surface drains in the direction of the slopes, as in ordinary wheat and
grass farming, it immediately bears good crops and is quite healthy.
There is no appearance of malarial poisoning. Such land, in the
older settlements of the country, is instantly recognized by the
elegant farmhouses and the big barns. The farmers occupying
such land may be known by their florid complexions and fine
forms. They are above the average size of men, bright and cheer-
ful. They mostly have large families, many members of which
live much beyond the average term of human life. The land
undergoes subdivision from generation to generation till the farms
become too small for further dividing, and then these healthy and
fertile farms send off a constant stream of emigrants to the large
cities, to the new countries. Nearly all the prosperous merchants,
physicians and lawyers, clergymen and philosophers, philanthro-

pists and sages, refer to these healthy farms as the homesteads of their fathers and grandfathers.

(6) In *the malarious district*, and on the farm of much malarious land, all this is reversed; the crops are so poor that there is no use for a big barn; the inhabitants are so poor that they are unable to build a good house, or even to keep their old hut in decent repair. The house has been built by some imprudent newcomer; he hardly raises any family, and probably he dies of the fever; one of his unfortunate children may inherit the place, to be sickly, poor, and wretched. Such farms are not much subdivided; the native population does not increase, and is not sufficient to keep up the occupation of the land; an occasional stranger becomes an owner, or two farms are united into one; and thus it goes on from generation to generation. The children are a sad sight,—sallow, dwarfish, deformed little creatures, a few of whom live to become very wretched men and women. Even the cattle, the very horses and cows, are dwarfish and worthless.

All this contrast comes of naturally gentle slopes, a moderate quantity of sand and gravel in the subsoil, and a sufficient elevation above the watercourses for good drainage, on the one hand; and, from the land being without one or the other of these advantages, on the other hand. The first of these pictures is seen in Byberry, Philadelphia, where nearly all the land is divided into very small farms, with elegant buildings, owned and occupied for six generations by people bearing the names that crossed the ocean with William Penn, and in similarly healthy districts of Bucks and Montgomery, and indeed nearly all over the country, except where overrun by cities and manufacturing villages. The second picture is seen in some parts of New Jersey; and in Pennsylvania it is seen in the southern part of Philadelphia, where the old township of Tinicum seems never to have had any barn at all, and but one farmhouse, and that never occupied by a family. The same picture of misery is seen in large districts of Delaware, Maryland, and Virginia, principally east of the main lines of railroad. All this flat country, except the pine barrens and the bluff banks of rivers, is full of misery, which the artist may occasionally admire for its picturesqueness. By careful study and systematic labor this may all be changed; this misery may be relieved; these ague-cakes and sallow skins, this sickness, poverty, and premature death.

(7) *Meadows and Valleys.*—It has been observed by Lind that

"*the most healthy countries in the world have unhealthy spots.*"
The country of natural subsoil drainage is no exception to this
rule.   The rains that water healthy hills wash down some of the
soil, and this settling in the hollows forms nearly level meadows
along the brooks and rivers.   The material thus deposited, the
soil of these meadows, is mostly clay, scarcely permeable to water.
The underdrain of these hills, in its natural channels, meets the
clay at the side of the meadow; it can go no further underground,
but it rises to the surface in elegant springs.   Such meadows produce no good crops, not even good hay; they produce a great

Fig. 4.

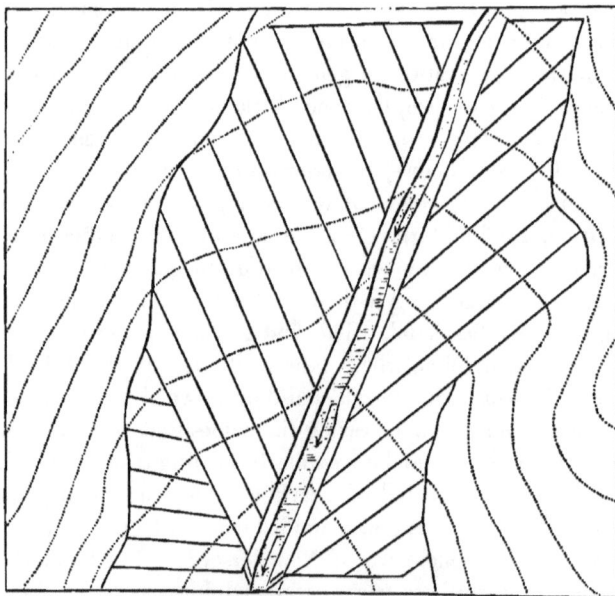

Drainage Map of the Marsh Meadow.

variety of sedges and bulrushes of very little value. These meadows
poison the air, causing malarial fevers, and thus they destroy the
health of all who live within the range of their influence.   The
remedy for this is simple enough; it involves some labor and
expense; it is merely underdraining, something equivalent to the
hole in the flower-pot.
    The drainage of meadows is most economically effected by a
system of underground drains, made with drain-pipes of brick-clay;
this is the cheapest material, and lasts for ages.   The first thing is
to survey the land; the surveyor, observing the neighboring lands

and hills, determines the lowest point of the meadow and measures the slopes. Between elegant sloping fields on each side there is a strip of *worthless marsh*, with a sluggish little stream through its middle; he draws a small map (Fig. 4), and, starting where the stream flows out, the lowest part, he traces dotted lines of equal elevation, each line being a foot higher than the preceding. The ground being very level, he must indicate lines for the drains nearly at right angles to these lines of equal elevation, in order to give the drains enough slope to make them work well. Commonly we may give all the drains enough fall by arranging them parallel with each other in two or three directions, as in the map. Thus we first trace a drain along each side, high enough on the slopes to catch all the springs; next, about twenty feet from our boundary lines, to drain our own land without interfering with our neighbor's; next, a drain on each side of the stream, nearly parallel to it, and thirty or forty feet distant; the rest of the land may have drains nearly parallel with each other, and as nearly perpendicular with the lines of equal elevation as possible, without too much complication of plans. In practice it seems sufficient to place the drains four feet deep and forty feet apart. Drains two feet deep have been found so inefficient that the work had to be done over again. Drains sixty feet apart did the work so poorly that it was necessary to make intermediate drains. Drains three feet deep must be nearer together, not more than twenty feet. A fall of six inches to the hundred feet seems quite sufficient for drains of this kind if the work is really well done, and there should be no part of the drain sloping the wrong way.

(8) In healthy countries we find other unhealthy spots, and consequent disease,—disease and death from very small marshes. Friedel mentions that, "in the Marine Hospital at Swinemünde, near Stetten, a very large day ward was used for convalescents. As soon as any man had been in this ward a few (?) days he got a bad attack of tertian ague. In no other ward did this occur, and the origin of the fever was a mystery, until . . . . . a large rain-cask, full of rotten leaves and brush, was found; this had over-flowed and formed a stagnant marsh, of *four to six feet square*, close to the doors and windows, ... which, on account of the hot weather, were kept open at night."—PARKES.

We have an account of another small marsh: "At Kingston (Jamaica) three young men, recently arrived from England, took up their residence in a large, airy house, on a place thought to be healthy. These young men were attacked in succession by fever; two died; the third had a severe attack of bilious remittent; he was removed, and eventually recovered. A large garden tank was found under the windows nearly filled with decayed vegetable matter; no one doubted that the malaria sprung from this."

Near our own homes we sometimes see a small marsh something like this though very much more abominable. Many of our country houses, and not a few in the cities, have no reasonable arrangement for getting rid of kitchen slops and garbage; the result is a *kitchen-midden* near the back door. This may possibly form an interesting subject of study for the ethnologists of the distant future; but at present it is disgusting and unhealthy; and it should disappear as soon as we can manage to educate our people up to a comprehension of the subject. Back yards and back streets, all over the cities, are constantly reported as nuisances by the sanitary inspectors.

We have another constant danger in the healthy country from frequent interference with the natural watercourses for manufacturing purposes. New mill-dams are made, ponds and marshes inconsiderately set up without care for the health and the lives of the people who are ignorantly dwelling near them; extensive public works, in the way of canals for drainage or navigation, have generally caused extensive epidemics in this way.

(9) *Heavy Land.*—In some places the earth to a considerable depth is composed of adhesive clay, such as is used in making bricks; it is almost impermeable to water, so that a shower of rain, instead of soaking away, must flow over the surface and thus keep it rather muddy; this is called heavy land. It sometimes occurs on the tops of pretty high hills, so that the highest land of a hillside farm is sometimes the most marshy and needs artificial drainage. The object in this case is accomplished in the same way as in the flat meadow; but we need not look for springs, and there is so much slope that there is no trouble in giving the drains enough fall without being quite so particular in giving them the exact direction of the slopes. But I am reminded that sometimes "experience teaches," and a few years ago I became interested in the condition of a small farm of this kind; about half of it was heavy land. *The first tenant* gave a very favorable account: "The place was well in with grass, and as I had work enough on my own place I took good care not to spoil the grass; the hay sold well and I had little to do, except to cure the hay and gather

the fruit for market. The back field had been in corn, and I knew that the land is too heavy there for potatoes, and so I put it in oats, to be followed in the fall with wheat and grass seeds. This succeeded very well; the oats was good, nothing to.brag of, but the wheat was about as handsome a growth as I ever saw, and it averaged fully forty bushels to the acre; the grass was about the same as on the rest of the place, perhaps two tons to the acre; this with hay at twenty dollars the ton does pretty well."

The second tenant was a sheep butcher, and I never saw him to hear what he had to say about it. His sheep pastured on the place; he ploughed a part of the front field, perhaps for turnips, and the water from the rest of the place, running across the ploughed field in heavy rains, washed out a deep gully, some part of it four or five feet deep. During his last year he ploughed nearly half the place and sowed it with winter grain, so as to keep the tenant of the next year from any profitable use of the land.

The third tenant, a market gardener, on account of the poor condition of the place, had it at a very low rent for the first year, but he did not prosper. He says: " I like the house and its situation very much, and my family like it. But the land is not good; some of it is right poor, and in these upper fields it is so heavy and so soggy that it hardly pays for manuring. I have tried and cannot get a good crop of vegetables on it; it is only good for grass. This old orchard where we are standing is very good. I have fixed the gully in the front field so that the washings have nearly filled it up; that front field is a splendid piece of ground, and it brings good crops of anything you choose to plant. But I can hardly make a living for my family here; there is not enough good ground, and I lose money when I pay wages for planting and tending crops on that heavy land."

The fourth tenant was a teamster, and his farming operations were not very successful. He says: " I cannot make out much at farming, and so I do hauling for the lumber yards and coal yards. We have had a good crop of potatoes on the front field and in the old orchard every time, but the land is so low and so wet that we never can get on the other fields in the spring early enough to plant anything properly. We tried to plant the back field with corn, but it was too wet; the horses could hardly go on it till June, and then it was so late that the corn did not grow well, and it did not have time to ripen, so that we did not have the quarter of a crop. The old fruit trees are very much in the way and the boys are all the time wanting to cut them down."

" But the trees must be useful to you ; when the other crops are poor the fruit, at any rate, must bring you something; and even if you have no sale for it fruit is not a bad thing to have in plenty where there is a family ? "

" Yes, that is so ; but they really are much in the way in plough-ing; some of them bear very little fruit, and some bear only half the time, every second year. You see that old pear tree, that Catharine pear ? I sold the fruit on it last year for thirty-eight dollars ($38); the buyer picked them himself and was right well pleased with his bargain. The children, may be because they heard the men talk about chopping, tried to cut down the next tree, just like it, and I wanted to give them a thrashing for fear they would get at it again, but my wife would not let me."

" May be that was best, for children can generally see whether you are pleased or not with what they do. Did you ever hear about George Washington, when he was a child, and his little hatchet ? "

" No, I never heard of it."

" Well, I have the whole story in a little book, and your daugh-ter would like to read it to them. You can tell them what a good thrashing they deserve for cutting that tree, only, may be, they did not know any better. But you must not allow any tree to be cut; you might be prosecuted for damages and sold out by the sheriff for allowing a few trees to be spoiled. That Catharine pear tree is about fifty years old, and is growing bigger and better every year; fifty such trees may stand on an acre, so that if the place were only covered with such trees, and if we could get them to bear every year, every single acre of the place would bring us nearly two thousand dollars a year without any work ; we would just walk out in the shade and let them pay us the thirty-eight dollars for each tree as they go to picking. I think that old tree is worth three hundred dollars ($300) just for its fruit. And shade is a good thing, good for horses, good for cattle, good for children, good for everybody ; the place would not be fit to live on without the trees."

"That is so, but I have hard work to make a living for my family here. I want to stay and will do the best I can. The back fields make hay and pasture without much work ; the front fields will give crops and fruit; and if hauling is good we will get along as well as common."

The fifth tenant, a practical farmer, had long been living on farms rented by the year. He could "make both ends meet" almost anywhere, but he never did much more. He said: "I am tired of big places, many hands, and heavy work; the little place suits me exactly if we can make a bargain. I prefer to work on shares, so that when crops fail I need not get into debt, and if the crops are good I get well paid for my work. I have stock enough for a larger place than this, so that the landlord need not put in anything for stock and tools." The lease was arranged to suit. The back fields produced hay and pasture, and the old gentleman did some profitable trading in horses. He did not get a very good crop of corn on the back fields, and one season he actually had to buy corn for his cattle. The fruit was good, but most of the trees were old, not good for much, and very much in the way in ploughing; the young trees not big enough to bear much. But the little place suited; the large family occupied the place reasonably well contented for nine years, till the old gentleman died. The buildings and trees were well preserved, with a general appearance of neatness and order, but in other respects the arrangement was not profitable.

(10) These reports are remarkably consistent: "Some found fault with the fruit trees, which were either too old or too young, or they were in the way, and would only bear half the time. All seemed to agree that the top of the hill is wet and heavy, and not fertile."

The sixth tenant has no family with him, his children being grown and married; other relatives live in Maine. He said: "I like the looks of this place; I would like to farm it. It may be poor, as you say, but I do not think it can be so very poor. There is not much grass, to be sure, but the cows I saw here last fall continued wandering over the bare fields, after the ground was frozen, till they had eaten out the very roots. At any rate some manure will set it right. I do not want any wages, only what I can get out of the ground. I need an advance of money to buy a stock of horses and tools. I want a liberal allowance for manure, and if the place is so dreadfully poor we must have a good deal of it. I just want to show you the crops that will grow here."

"But they say that the back fields are so low and wet that

nothing hardly will grow; and no matter how much you manure, nothing grows well."

FIG. 5.

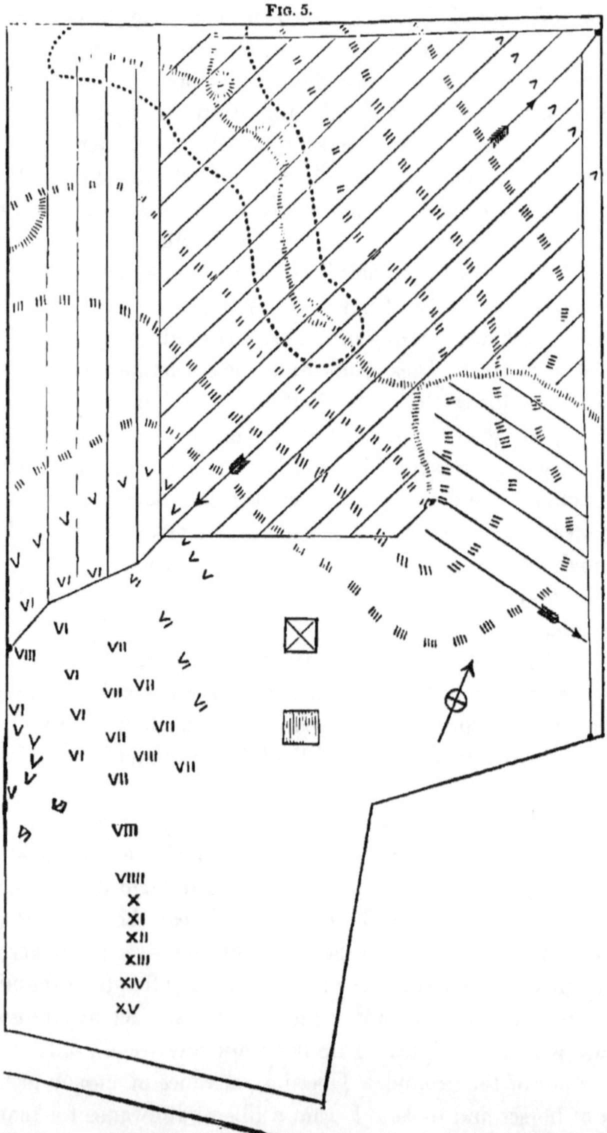

The Drainage Map for Heavy Ground. The surveyor, for the sake of variety, sometimes begins at the top of the hill and works down.

"How can it be low, right on top of the hill? How can it be

wet? it is a regular slope, and I see no place where a drop of water can stand on it."

" Yes, it is a nice slope; but the soil is clay, the same they make bricks of. You have seen men in the brickyards, in the beginning of winter, digging up the clay into heaps, to temper, as they say. It gets several freezings and thawings, and in the spring it is as soft and sticky as so much tar; and it remains soft and sticky till they mould it into bricks and spread them out to dry. How can you take horses on such land to plough for corn early in the spring?"

" I can find a time to plough it; only get me a manure-wagon and let me try. You will soon see something growing."

This is evidently the man for the place, with enthusiasm for the work. He received his first lesson in his attempt to raise a crop of corn. He ploughed and he planted; the corn sprouted and it grew with weak stalks, many of them broken down by the wind; the leaves were yellow in stripes—very pretty. As for corn, there was not much—only a few nubbins.

" Well, I told you so."

" I know that; but what can we do? We shall have no corn." . . .

" I have ordered five thousand feet of drain-pipe, to be delivered in August, so that everything may be ready for digging trenches as soon as the corn is out of the way. You may begin by cutting rows of corn across the field where the trenches are to be. Here is the map all ready (Fig. 5). The drains may continue further up the slope, so as to make a field of three or four acres. Let us try for another crop of corn, but you need not forget the manure-wagon."

The next crop of corn was a success, with stout stalks, dark-green leaves, handsome large ears, and a very large crop of corn. Our present plan is to drain three or four acres each autumn, so long as we see the land to improve under the operation.

(11) We had much to learn in regard to the manner of doing this work. The first set of men employed to dig trenches made them so wide, finished them so neatly, and worked so deliberately that it soon became evident that it would cost more to dig the trenches than the land is worth. Discharging the workmen and employing others mended the matter; but still the cost was so great as to make the work impracticable—it would not pay. A clay-digger from a brickyard, out of work on account of the dull times, came along and proposed to finish the job himself for fifty dollars ($50). He made his trenches about half as wide as

the others; he did not waste much time in trimming the sides; he finished the work in good time, and he earned fair wages.

Fig. 6.   Fig. 7.   Fig. 8.

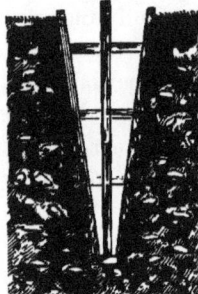

Form of the Trench.   Frame to Regulate the Size and   Propping the Trench in very
                         Form of the Trench.   Soft Ground.

To regulate the slope of the bottom of the trenches it is sometimes advised to use the mason's level, with some special addi-

Fig. 9.

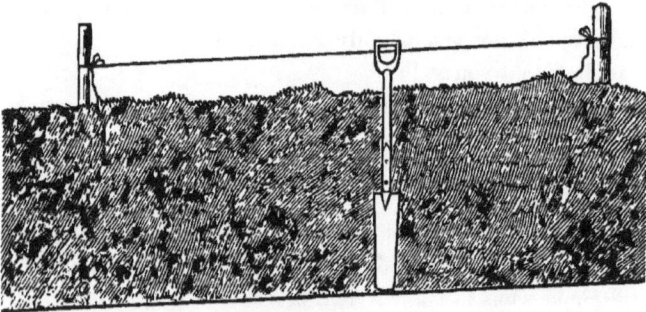

The Slope is regulated by stretching a line between two stakes and measuring with the spade.

tions, or to wait for water to run in the trench, so as to indicate the proper level. We find these things inconvenient; especially

Fig. 10.

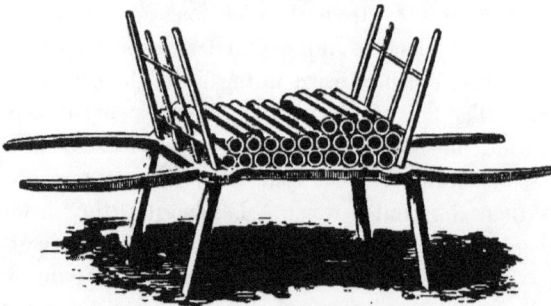

The Hand-barrow.

the water made a troublesome amount of mud and caused the sides

to fall in, so that the work had to be done over again. Our plan
is to drive two stakes near the trench, at some distance from each
other, and to secure a line to them with the proper slope, so that
the workman with the handle of his spade can measure the depth
from the string. When the land is ordinarily moist the water

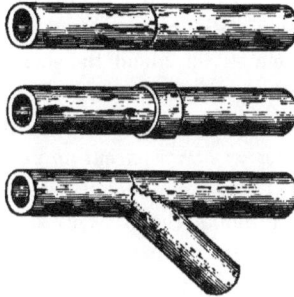

Fig. 11.

The pipes are simply placed end to end in contact, or they are united with collars, and the
best Y joints are made by chipping with a very light hammer.

filters pretty rapidly into the trenches, so that we find it advanta-
geous to put in the pipes close up, and to cover them with enough
earth to secure them in position before the water has time to
get in.

The pipes are of various shapes. We prefer the simple round
form, so that, no matter which side is up, the pipes always fit.
The common pipes are, in large proportions, curved, warped in
the kiln. They need not be altogether rejected on this account,

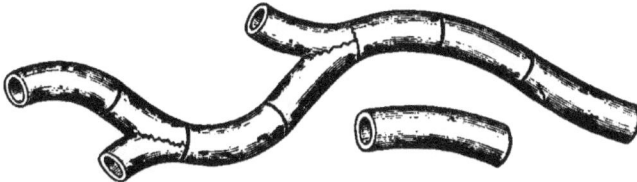

Fig. 12.

Drain-pipes Warped in the Kiln, are very useful for making easy curves, and for connection
of lateral drains.

as water does not refuse to flow very well through regularly curved
channels. There is positive advantage in having a portion of them
thus curved, so as to make easy curves, instead of angles, in changing the direction of drains.

The size of the drain-pipe is a matter of some importance. It
should be large enough to carry all the water, and any larger size
is a waste of material and a needless expense.

"Mr. Thomas Hammond, of Penhurst, Kent, uses no other size for parallel drains than one-inch pipe, having commenced with 1⅓ inch; and the opinion of all the farmers who have tried them . . . . is that one inch diameter is abundantly large. A lot of nine acres sown with wheat was observed by the writer 36 hours after . . . . a heavy rain of 12 hours. The length of each drain was 700 feet; each drain emptied separately into a running stream, so that the discharge was distinctly visible. . . . One which tapped a small spring gave a stream about the size of a tobacco-pipe (?); the run from the others did not exceed the size of a wheat straw. The greatest flow at any time had not . . . more than half filled the pipes. The water was transparent and clear."

We have not tried these one-inch pipes, as our manufacturers are not accustomed to make them. Our smallest pipe, as well as our largest, is two inches. A main drain of this size discharges the drainage of seven acres, and during three years it has never had water enough to use its full capacity. Therefore our experience accords nearly enough with that of Mr. Hammond in this quotation, and with Mr. Waring's table of sizes. We have accordingly arranged the following table, assuming that the velocity of flow increases regularly with the cube of the linear dimensions.

For 1 acre or less use	1 inch pipe, with	2 inch collars.
2 acres "	1¼ "	2¼ "
2½ " "	1¼ "	2½ "
3½ " "	1½ "	3 "
5¼ " "	1¾ "	without collars.
8 " "	2 "	" "
15⅝ " "	2½ "	" "
27 " "	3 "	" "
54 " " 2	3 "	" "
64 " "	4 "	" "
125 " "	5 "	" "
128 " " 2	4 "	" "
192 " " 3	4 "	" "
250 " " 2	5 "	" "

There is no real use for all these sizes; the 1¼-inch pipe can hardly cost more than the 1-inch, and may take its place. These small pipes, they say, should always have collars, as otherwise it may be difficult to keep the ends together with certainty, and the 2 or 2½-inch pipe in short sections makes the collars. 2-inch pipe, or 2½, will do for main-draining, till we come to a drain to carry the water of about sixteen (16) acres. Thus we need but three sizes,—1, 2, and 3 inches, or 1¼, 2½, and 4,—for we are not likely to meet with a case in which it may seem convenient to drain more than sixty-four (64) acres through one pipe. There is positive disadvantage in having the drains too large, for the only obstruction is by fine silt, which can lodge only in a slow current, and must certainly be carried through these small tubes by the rapid current produced by a rain-storm.

The circular form of drain-pipe is the best. The horseshoe tile, open below, is better than the old drain of stones or brush, and it is sometimes used with a board floor; but this is absurdly expensive. The sole-tile is a pipe with a flat surface below, but this flat surface is inconvenient, and of no use; elliptic and oval forms have been tried, but they have no advantage to compensate for the additional labor of placing them.

The plough is used to start the trenches, and we need a few peculiar tools; the curved clay-spade of the brickyards is

Pipes are made of various forms.

useful, but it is much too broad for the bottom of the trench; a much narrower spade is needed, about four inches wide, not tapering much, or not at all, and curved laterally like the brickyard spade; a common flat shovel, with long handle; a much narrower shovel, curved laterally; a scoop, with long handle, to dress and clean out the bottom of the trench; a horse-shovel, or scoop, to fill the trenches by horse-power, always, of course, taking care that the pipes are secured by a light covering with the shovel, so as to run no risk of being displaced or broken by a falling pebble. There is no need of ramming; the frosts of winter and the rains of early spring pack and puddle the earth about the pipes, so that as soon as the land is ploughed the work is all out of sight.

The *outlets* of these drains are generally described as involving some care and expense, with solid masonry, and gratings fine enough to exclude toads and mice. So far we see no use for these things. The last six or eight feet of the drain is made of glazed and hard-burned pipe (terra-cotta), so as to resist frost or the trampling of cattle. The drain comes near the surface, where there is a convenient slope under a hedge or fence, or to the bottom of a surface-drain by the side of a road. The pipe terminates in a basin about a foot wide, as deep as the diameter of the pipe, and it is soon

covered and concealed by grass spreading from the sides. This is
the outlet. In ordinary rains no water reaches it; after a heavy
rain a little water runs for a day or two; in the spring, from the
melting snow, there is a full stream for a week or two. The water
is perfectly limpid and colorless, apparently pure, notwithstanding

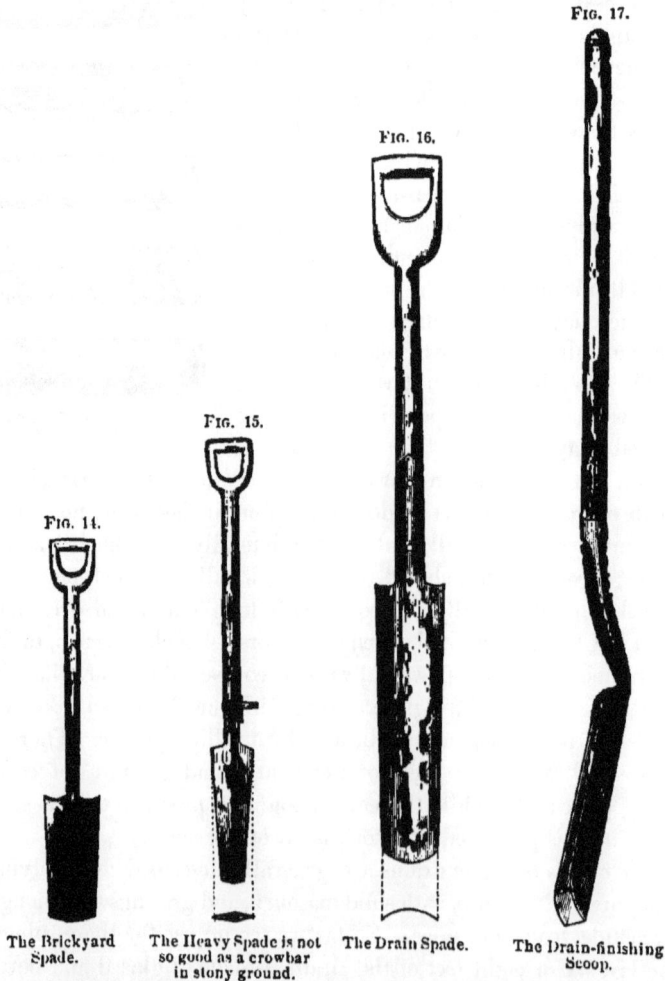

FIG. 17.

FIG. 16.

FIG. 15.

FIG. 14.

The Brickyard
Spade.

The Heavy Spade is not
so good as a crowbar
in stony ground.

The Drain Spade.

The Drain-finishing
Scoop.

the ten or twelve tons of manure to the acre. If this drainage-water
should become troublesome it can be discharged into a deep sink,
in accordance with the plans represented in the margin (Fig. 18).

(12) This heavy land, whether underdrained or not, when
regularly and sufficiently sloped, must be healthy. The small

farm here discussed has been occupied continuously as a separate farm for sixty-five (65) years, hardly ever by a family of less than six (6) persons, and much of the time by eleven or twelve—average perhaps eight. The deaths on the place have been seven, which is 13.46 per 1000 per annum. We are conscious that two of these deaths were by malarial fever, which may have been caused

Fig. 16.

Sinks to discharge water into permeable deep strata.

by a disgusting mud-heap, a little marsh or kitchen midden, near the back door (§ 8), for we know that such a nuisance at one time existed.

(13) *Sandy Loam.*—A large proportion of the malarious land of the United States is sandy loam, so flat and so nearly on a level with the ocean that it is very imperfectly drained. There is some of it in New Jersey, Delaware, and Maryland, and in Virginia; it includes nearly all the State east of Fredericksburg and Richmond. Some of the best farm-houses in this district are built of bricks imported from England by the early settlers. It still continues in farms of a thousand acres each, or even larger. The poor owner does not know how he can do any better, and he really has no means to improve his land. The healthy lands of this region are the bluff banks of rivers, the pine forests, and some sea-coast islands. This border of level sandy land extends along our coast from New Jersey to Texas, varying in width from twenty (20) miles to 150 miles. For half its breadth there is hardly any of it thirty feet above the level of the ocean. With an average slope of less than a foot to the mile it is not well drained, and it would seem nearly impossible to drain it; but even this desperate strip of land may be improved. The very worst of it can be cultivated in rice, but no one should think of fixing a dwelling near the rice-fields; the workmen can live in healthy villages at a safe distance,

and, if necessary, while planting and harvesting the rice crop they can sleep in barges anchored in the middle of the streams, merely landing for work in the daytime; for it has been abundantly proved that there is no danger of malarial poisoning in the daytime, and very little danger in a boat anchored half a mile from the malarious shore, even at night. And in this desperate strip of land there are healthy spots,—the bluff banks of rivers and the pine groves,— and in these spots villages of prudent people will be healthy.

(14) The western half of this sandy plain is considerably more elevated, some of it even reaching two hundred feet above the level of the sea. This is the region of pine barrens; it is too level and too much under water for cultivation without some system of artificial drainage. The country is sparsely inhabited by an unhappy malaria-poisoned people. The lumbermen, while in the forest, are healthy, but as soon as they clear a field for cultivation they have the fever, they die, or they become helpless invalids, and the weeds, the bushes, and the forest again gain possession. An attempt has been made to drain by open ditches along the roads, and in some places ditches are made to serve for boundary lines and fences. But there are not enough ditches to drain the land; they breed terrific swarms of mosquitos, and they rapidly fill with leaves and grass, so as to need a constant cleaning out.

There are several ways to improve this land. In the first place we must insist on the fact that but for the growth of pine trees this country is deadly from malaria, and hence that the people must live in villages, in the healthy spots protected by groves of pine trees. They can walk to their fields in the morning and return in the evening; an occasional foolhardy one or a drunkard will remain in the fields at night, and he will have the fever in about fourteen days afterwards and will probably die, so that the rest will be admonished to keep good hours; thus we may secure health for the workmen. Next, instead of the few open ditches, which only half do the work and are more expensive, we must have drain-pipe and have it carefully laid from three to ten feet deep; we must have five times as many such drains as there are now of open ditches. These will be out of the way and out of sight; they will carry off all the surplus water, will breed no mosquitos, will need no repairs or cleaning out.

With these arrangements this land may be made productive, and when it becomes healthy the foolhardy will soon find it out and make the fact known. The mere swamps along sluggish streams, though not habitable, may be made valuable as rice fields.

(15) *Pine Forests.*—There is something curious, something astonishing, about the salubrity of the pine barrens of this region, which, of course, are not barren at all. These immense tracts of flat land are occupied almost exclusively by pine trees. The trees grow close together so as to destroy side branches, and thus we have tall timber trees, and the earth is covered some inches deep with dead branches and pine leaves. The ground is quite dry, the water disappearing very rapidly after a rain. These forests supply the markets of the world with tar, timber, and turpentine. But the interesting matter to us is that they are quite healthy,—

Fig. 19.

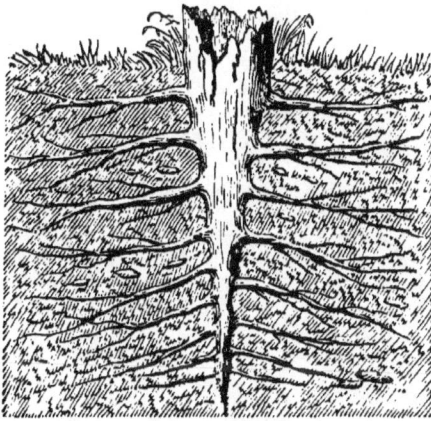

The Pine Stump—as it decays makes a good drain.

quite clear of malarial fevers. Some attempts have been made to account for this immunity, this good health. It has been suggested that turpentine is wholesome, and that it neutralizes in some way this poison, and, although there may be some truth in this notion, there is no exemption from fever about the turpentine stores except in the forest. The thick carpeting of pine leaves has been supposed to prevent the malarial poison from rising, and it doubtlessly is beneficial in this way.

It seems to me that the pine trees act principally by draining the land, and this they do by the peculiar manner in which the roots decay. When the trees of the forest get too thick to thrive the weaker die out and decay, root and branch; but the roots decay much more rapidly and more completely; thus the whole situation of the pine stump becomes a very effective sink. The roots

3

near the surface form a series of collecting drains, and the deep
roots are distributing channels by which the water rapidly sinks to
its lowest level. These stump-holes, with the cavities left by the
large roots, are very nice dens for coons and foxes. There are
some holes of this kind in the pine grove of the Naval Hospital
at Norfolk, Va. These holes are four or five feet in diameter to
an apparent bottom of twigs and leaves. I asked many questions
and could only elicit the conjecture that they might have been
made by the diggers for Kidd's money. This explanation would
not do as there was no visible pile of dirt. A surface drain was
opened into one of these holes with the expectation of filling it up,
and after the next heavy shower I saw a stream of water, eighteen

FIG. 20.

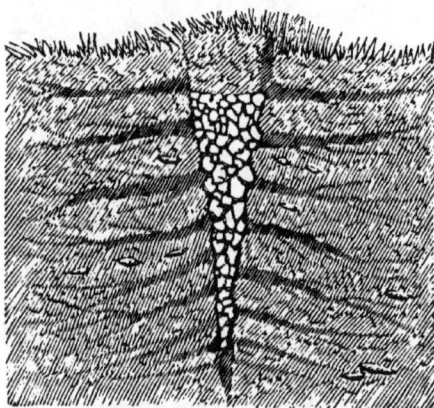

Pine Stump Drain Improved.

inches wide and two inches deep, running directly into the hole
and disappearing—a small Niagara. An explanation eventually
came by accident. Trying my cane on an old stump it penetrated
easily in any direction; the strongest part of the stump was its
bark, the rest was black dust. All the men about the place seem
to know that the pine stumps decay in this way, and very quickly,
even to the ends of the smallest roots; thus the matter is plain
enough. These stump-holes might be utilized as permanent drains
and sinks by merely filling them with gravel or even good chips,
except about two feet at the top, and finishing with earth like the
sinks already represented (Fig. 18, § 11).

This action of pine stumps in draining land begins when the
trees are quite small. The first year from the seeds the trees are
a foot high; the second year they are four feet; the third year

they are eight feet, and more than half that started are dead and some of them decayed. Every year while the forest lasts there are trees thus dying and decaying with their deep tap roots. It seems easy then to make a healthy place for a village almost anywhere, as we have only to scatter a few pine seeds on any suitable soil and to wait three or four years. The Naval Hospital at Norfolk is separated from very malarious fields by a pine grove less than a thousand feet wide, and this is found to be quite sufficient.

The *eucalyptus* of Australia is just now exciting much interest as a malaria-preventing tree, but as yet there is no evidence of its being so good as our own Southern pines, which constitute these barrens. However, it is an elegant large tree, and there is no danger of too many such good things; and, besides, this tree may be suitable for situations where the pines do not flourish.

(16) *Flat Rock.*—Sometimes we find marshy land on a nearly level surface of flat rock. With ordinary tools not much is to be done, but the same general plan is adopted. The rock may be limestone, not very hard, so that with picks and crowbars the trenches are opened, not very deep or very far apart. Instead of drain-pipes in the bottom of the trenches we arrange fragments of the rock; the earth is filled in and levelled, making a French drain (Fig. 3, § 3), and thus the work is completed. This kind of draining is practiced at Sandusky, Ohio, and thus poor wheat land worth thirty dollars the acre becomes worth four hundred dollars for vineyards, and the country is becoming healthy.

(17) *Bank Meadows.*—In draining land subject to a regular overflow of tide, the water is collected into a pool near a sluice-gate and discharged into the river at low tide. These bank meadows along the Delaware, and probably everywhere, are dangerously malarious, destroying the health and the lives of nearly every person who attempts to dwell near them.

(18) It might be inferred that the malarial poison that kills so many thousands is the same that prevents our plants from thriving, but this would be a mistake. The marshy meadows of New England produce nothing but meadow hay, good to pack crockery, but malarial fevers of the severer types are scarcely seen in New England. The land occupied by the Panama Railroad has an abundance of marshy places, but not at all in proportion to the deadly character of the climate. The lives of more than eighty thousand (80,000) men were destroyed by malaria while employed

in building this railroad. One man died for every yard of the track, 1760 died for each mile, and in round numbers 81,000 lives were destroyed in building forty-six (46) miles of railroad, and it is to be feared that five times as many will be destroyed in a contemplated enterprise of building a canal in the same country.

There appears to be at least three distinct species of marsh poison removable by drainage: (a) There is the poison that interferes with vegetation, so that some meadows produce sedges and rushes instead of grass; (b) there is some influence which determines that typhoid fever is more prevalent on flat or marshy land, though not confined to such situations; (c) and most important of all is malaria, the poison of the periodic fevers, killing its hecatombs, say ten thousand (10,000) victims every day of the year, nearly one death for every tick of the clock, and keeping a very large portion of the earth's surface in the condition of a howling wilderness.

But possibly the marshy soil does not generate either of these poisons after all. Grass does not grow well in a marshy meadow, but this division of plants finds its appropriate nourishment on dry land; that it does not grow in water may be from the absence of something necessary for its nutrition. The mechanical difference between light soil and marshy soil seems somewhat similar to the conditions of a wet sponge just lifted from the water and of the same sponge partly dried by pressure in the hand. In the one case we have a continuous body of water with mingled particles of solid matter, without air-space; in the other condition there is moist solid matter, with intervening air-space. The air thus mechanically present in the light soil must expand and contract with changes of temperature; and consequently in the usual changes from day to night it must be drawn in and expelled, in a way comparable to the respiration of animals. Now it has been demonstrated that the atmosphere is everywhere full of minute objects, some of them living organisms, of many kinds, millions of millions of individuals—a typhoid fever germ, a yellow fever germ, a malarial germ, a scarlatina and a variola germ may be among them. When the light soil thus takes a long breath, myriads of these objects, these germs, are inhaled, and adhere to the moist interior of the air-spaces, and are thus consumed in this grand laboratory of nature. The marshy soil having no air-spaces to absorb and destroy the germs, they must remain at the surface or continue to float in the atmosphere. Whether this fanciful theory is true or whether it is mere fancy is comparatively unimportant to us; the facts on

which it is based are true, and are exceedingly important. Men
and women are healthy, happy, and prosperous dwelling on this
light soil or on well-drained land; they are sick and wretched,
miserable and dying when they undertake to dwell on the marshy
land—they cannot live there.

(19) I hope to be excused for a little more persistence on this
important subject. In all parts of New England hundreds of
people are dying of typhoid fever; a large tract of the city of
Boston is now building on made-land nearly as flat as the prairie
around Chicago, and in a few years it will probably have to be
regulated and rebuilt to get rid of the pestilence. From Maine to
Pennsylvania there are undrained fields nearly as bad. All over
the country further south, but principally in the Mississippi Val-
ley and in the flat country bordering the ocean, the undrained
land is infectious with intermittent fever and the other malarial
pestilences to such an extent as to destroy many thousands of lives
every year; so that in spite of constant immigration extensive tracts
of country continue as sparsely peopled as ever, with very unhealthy,
very unhappy people. Some part of this desert is the most fertile
land in our country, and the most easily cultivated but for the
failure of health among those who undertake to occupy it. In
1843 I became acquainted with some men who had purchased
tracts of this land remarkably cheap—the crops of the first season
more than paid for the land; but before Christmas one or more of
each family were dead, and the rest were so broken down in health
that no other good crop was ever raised by them. The migrations
to this pestilential region are constant, so that on visiting the same
region a quarter of a century later, in 1868, the same desert was
found, the same paradise of mosquitos, the same style of man,
with apparently the same gun, the same ox, same cart, same children,—and the same kind of new immigrants as hopeful as ever
over their first crop of peas and cucumbers, and older settlers
bemoaning the loss of relatives. It is best not to hint to these
poor people that the country is not the most healthy in the world,
or the man that leads the ox might shoot you for abusing his
country; discretion is much better than either dogmatism or argument on this subject till you get well beyond musket range. He
knows very well that they have no sickness but the chills, and
"any fool knows" that is no sign of an unhealthy country, for
people have the chills everywhere. These people seem to think a
good shake every second day an evidence of good health; and in

this they are probably so far in the right, that a person having a regular tertian ague may be less likely to suffer from the more fatal forms of malarial disease. (See frontispiece.)

It would be well if the migrations to these beautiful and fertile fields could cease until it can be undertaken with a fair appreciation of the difficulties; but it will not cease, it cannot be stopped, and any great change for the better must come by improving the sanitary condition of the country. I am confident that it will eventually become the happy home of millions of prosperous people, and groves of long-leaf pine (*Pinus tæda*) will protect and adorn the villages.

Thus the unhealthy country is known by the geological character and conformation of the land; by the repulsive sallow look, slim bodies, and shabby clothes of the people; by the poorness and small size of the cattle; by the fantastically primitive farm carriages; by the absence of capacious barns. The remedy is drainage. The *one great thing* would be a grand canal, similar to that of China, to connect Boston harbor with the Rio Grande—inland navigation half way between the hills and the ocean; and as soon as we may be conveniently annexed to Mexico it may be continued to the Coätscualcos, so as to have a double outlet at the south by way of the Tehuantepec Canal.

The drainage of the house and the garden is nearly as important as that of the fields, and will form the subject of the next chapter.

CHAPTER II.

THE DRAINAGE OF THE FARMHOUSE AND OF THE VILLAGE.

(20) There is some care to be used in the selection of building sites, and the plans must be suited to the situation, to the character and conformation of the land. A favorite situation is on a slope of high land, and this as the rule is certainly the best, as, other things being equal, it gives the best natural drainage. But this needs examination, and if the soil is of pure clay the building lot needs the same sort of draining that we have indicated for the fields. Sometimes there is a seam of sand or gravel, and if this is of the proper depth it drains the foundations nicely; but it may be so situated as to bring a portion of the drainage of the hillside

into the cellar walls. This of course makes the cellar wet every
time there is a heavy rain. The common remedy for this is to
make a drain from one corner of the cellar down the hill, thus receiving water into one side of the cellar and draining it off at the
other side. This arrangement, with a stream of water running

Fig. 21.

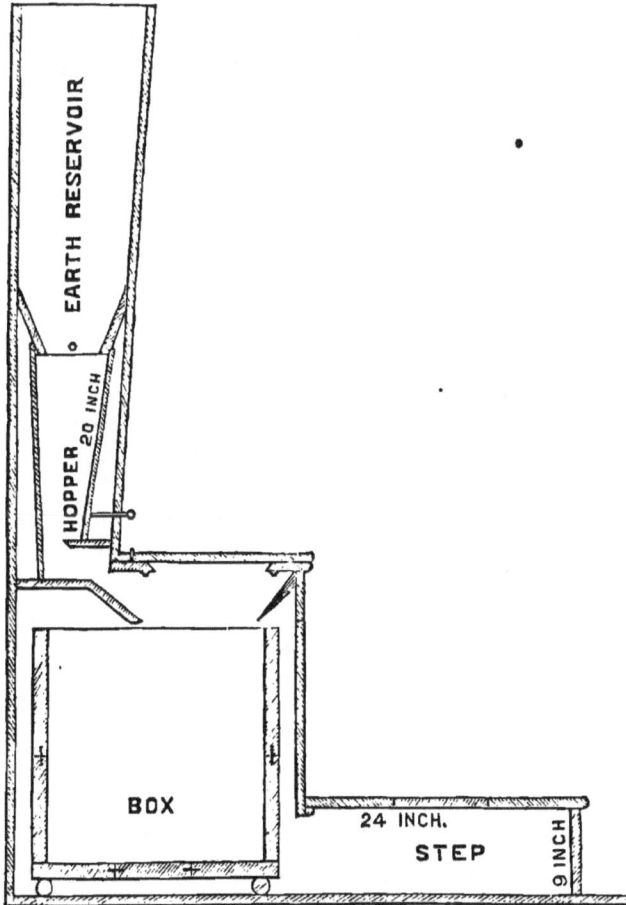

The Earth Closet.

through the cellar, is not very nice. An effective remedy, perhaps
the best plan, is to dig a trench a little deeper than the cellar on
the up-hill side, about ten feet from the walls, continuing it down
the hill, and to lay a drain of common pipe. The trench should
be nearly filled with gravel or sand, so that any intercepted channel
may continue to find its way down into the drain.

If the water-supply is by wells, in a light or gravelly soil, there can be no dirt-vault anywhere near without contaminating the water. Even in soil of the stiffest clay there is danger of the same nuisance. The vaults are sometimes so deep as to be really wells, the water mingling freely through the channels on the underlying rock. It is not hard to understand that typhoid poison may and does travel long distances in these deep underground channels. The safest way seems to be to mingle the dirt with the soil of the garden, there to decay and make the plants grow. The conveniences may be changed about from time to time, so that no particular part need become excessively contaminated. "The earth closet" is very convenient for this purpose (Fig. 21). The excrement being mingled with dry earth, or coal-ashes, or plaster of Paris, is no longer offensive, and it may be carted to the fields and spread like any other compost.

(21) Without any very great expense country houses sometimes have the usual city conveniences, with water-supply and bathing fixtures. The rain-water from the roof is conveyed into a tank near the top of the house; the excess in rainy weather runs into a cistern in the earth, to be pumped up if necessary in dry weather. If a fire is kept up all night, a hot-water tank, a circulation boiler, is so placed as to prevent freezing. The only real difficulty in keeping up this arrangement is the habit of wasting water, allowing it to run to waste without thought or care; for nothing seems more annoying to a man or a boy than to be called several times a day to pump a little water into the tank. It is very light work, but it is a most vexatious interruption of either work or play.

The fluid waste of these conveniences is carried off by a special arrangement, which disposes of the kitchen slops at the same time, by using them to irrigate an adjoining garden. The fluid is confined at first in iron drain-pipe, joints made tight by lead well calked; next in glazed earthen pipe (terra-cotta) with water-tight joints to a trap arranged to stop all solid matter. The water from the kitchen sink, having already passed through a similar trap to stop its grease and some dirt, increases the current, and the whole now at a sufficient distance (100 feet) from the house is turned into a channel of drain-pipe, with loose joints, about a foot deep (see Fig. 23). It thus mingles with the soil and makes the grass grow. This arrangement is easily made in a suitable situation, and requires very little care. But I will relate an incident, something that happened to myself.

(22) A few years ago I visited a farmhouse with a view to improvement. The diagram (Fig. 22) is intended to indicate the state of things—the usual condition on such occasions. There was a heap of mud and garbage—a kitchen midden—near the pump, evidently draining into the well. The farmer's wife urgently requested us to taste the water and to say whether it was fit for use; but we were not in the least thirsty. The mason who was to make the repairs was induced to express himself quite freely :

Fig. 22.

The farmhouse as commonly found when occupied by tenants, not the owners: (a) drippings from the eaves, cut holes partly filled with gravel, from which water filters into (b) the cellar; and (c) a heap of garbage, with kitchen slops and soap-suds, drains into the well.

" It seems that the water is nasty and not enough of it; the water on this place must be naturally good,—on this kind of land it is always good. Perhaps it would be a good thing to deepen the well ?"

" Yes, we have decided to deepen the well, but what is to be done with the mud and dirt ?"

"That is an easy matter; you can haul water in casks for the use of these people so as to relieve them of this nastiness; the water of this well will do to mix mortar till the well-digger comes. But well-water is a little hard for washing clothes. I would like to begin by building a small cistern about eight feet in diameter, to hold some of the rain-water from the roof, two or three thousand gallons?" . . . .

"Very well, we will have the cistern."

"There is no other doubtful question. You see the water from the roof has cut these holes about the cellar walls. I will clear out all these holes, mend the walls, level up and puddle with clay from the cistern a little higher than the general level, so that the slope may carry off any little water. If necessary we could pave the surface so as to make sure that it would not wash out again; but this is not worth the trouble, for if the gutters and spouts are kept in order it will never wash out. I wonder these poor people are not all dead; the cellar must often be half-full of water; and then the nastiness they have been drinking."

"What shall we do with the well and the pump?"

"The well-digger will be here in two or three days, and in the meantime we will not disturb the dirt. He will remove the wooden platform; he will place his windlass; he will lift out the pump, remove the entire wall, deepen the well, and wall it up again to within four or five feet of the top. The pump-maker will put in a suitable pump; the masons will finish the wall, the upper five feet in hydraulic cement, narrowing as they finish, so as to leave an opening only large enough for the pump and a man-hole. The whole of this upper part will be as tight as any bottle. We will select some earth from the bottom of the well, or from the cistern, to fill in and puddle up to the top. The material that comes of deepening the well must be carted away, and the present mud-heap will be so mixed up that it will be nowhere. The ground about the well will settle for some time—some years; but I can arrange the slope of the pavement so that with all the settling there will be slope enough to carry off the waste water. Some of the pavement and a paved gutter, twenty or thirty feet long, must be laid in hydraulic cement. Beyond that distance you can have a wooden trough across the garden, to leak and water plants as it goes. It will be nearly as it was when the house was new, but there will be hard brick and cement for thirty feet, instead of wood. With decent care it will be good for a hundred years."

This plan was substantially adopted. But the leaky wooden trough was omitted, and its place was supplied with an irrigating channel of terra-cotta pipe, about a foot deep, nearly if not quite in accordance with the plan described by Mr. Waring in the *Report of the Bureau of Agriculture* for 1871. We omitted the ventilating arrangement and cemented only the lower half of the circumference of the pipes, so as to ventilate into the soil and to receive drainage on occasions of heavy rain. The grease trap and the dirt traps are of plank, instead of bricks. This arrangement has been in use for several years without repairs or any care, except removing mud from the inlet occasionally, and cleaning out the traps.

(23) The situation of this house is excellent; its condition was disgusting. A young girl had recently died in the house. The masons wondered that they were not all dead. The heap of garbage was higher and wetter and nastier than either of the small marshes described on a preceding page as the cause of fatal malarial fevers (§ 8).

(24) The pestilence that seems especially to attach to such contamination of the water-supply is *typhoid fever*. In 1843, in the days of stage-coaches, a young man who had been to New York to purchase a supply of merchandise, was on his way home in the mail-coach, swinging about and dozing as the wheels bounded

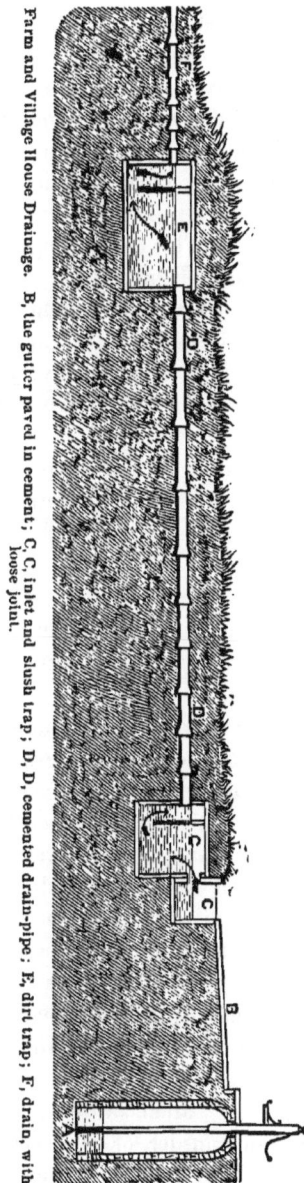

Farm and Village House Drainage. B, the gutter paved in cement; C, C, inlet and slush trap; D, D, cemented drain-pipe; E, dirt trap; F, drain, with house joint.

Fig. 23.

over the rough road. He became very tired; he was sick, and he
could go no further. The coach left him at a little wayside tavern
in the little village of New Boston, Erie County, N. Y. In a few
days he died; others died in pretty rapid succession. Of forty-
three (43) persons in the village, twenty-eight (28) were sick, and
ten (10) died. Three families escaped the pestilence; two of these
families lived a little way out, and had their water from a brook;
all the rest had been in the habit of getting their water from the
tavern pump. One family in the midst of the village had a quarrel
with the tavern-keeper and would have nothing to do with him or
his pump. This quarrelling family were accused of poisoning the
well. Austin Flint went to the village and investigated the matter.
The families who had been so sorely afflicted with disease and
death suffered from an expensive lawsuit and paid a fine for
slander. This happened a very few years after Louis, in Paris, had
demonstrated the existence of typhoid fever as a specific disease,
and after Gerhard and Pennocke, in Philadelphia, had taught us
the same thing. This is perhaps the first hint on record of the
possibility of typhoid infection being conveyed by water.

(25) The reports of such incidents are common enough since
attention has been called to the matter. Epidemics of typhoid
fever have been propagated by milk to whole rows of city houses,
and it is suspected that many cases of this "mysterious providence"
have been caused by watering the milk, by "milking the cow with
an iron tail."

(26) A few years ago, at one of the Croton reservoirs, there
was established a boarding-house replete with sanitary appliances
and elegancies. It was advertised as a *sanitarium*. People sometimes pay extravagantly for such care of their health. The house was
abundantly supplied with water by a pipe from the lake and stored
in a cistern, so that a small pipe kept on hand enough for any
probable emergency; another cistern was built so as to have water
convenient for the steam-engine, and another for the laundry, all
connected by small pipes, very convenient. There was some
trouble in disposing of the sewage from water-closets, and the
unhappy thought reached the sharp intellect of one of the propri-
etors that a disused cistern might be adapted to this purpose. The
consequent "Mahopac disaster," mysterious providence, dirty trick,
epidemic, murder, or whatever people may choose to call it, caused
about fifty (50) deaths of guests of the house by typhoid fever, and
a large amount of life-long decrepitude and suffering. The business of the *sanitarium* was ruined for the season.—*Sanitarian*.

CHAPTER III.

THE DRAINAGE OF CITIES.

(27) "Paris throws five millions a year into the sea, and this without metaphor. How? In what manner? Day and night. With what object? None. With what thought? Without thinking. For what return? For nothing. By means of what organ? By means of its intestine—its sewer.

"Thanks to human fertilization, the earth in China is still as young as in the days of Abraham. Chinese wheat yields a hundred and twenty fold. There is no guano comparable in fertilizing power to the detritus of a capital. . . . .

"We fit out convoys of ships at great expense to gather up at the South Pole the droppings of petrels and penguins, and the incalculable element of wealth which we have under our hand we send to the sea. All the human and all the animal manure which the world loses, restored to the land instead of being thrown into the water, would suffice to nourish the world.

"These heaps of garbage at the corner of the stone blocks; these tumbrils of mire jolting through the streets at night; these subterranean streams of fetid slime, which the pavement hides from you—do you know what all this is? It is the flowering meadow, it is the green grass, it is marjoram and thyme and sage; it is game, it is cattle, it is the satisfied low of huge oxen in the evening; it is perfumed hay, it is golden grain, it is bread on your table, it is health, it is joy, it is life. Thus wills that mysterious creation, which is transformation on earth, transfiguration in heaven. . . . .

"Imitate Paris and you ruin yourself. Moreover, in this immemorial waste, Paris herself imitates. These surprising absurdities are not new; there is no young folly in this; the ancients acted like the moderns. 'The Cloacæ of Rome,' says Liebig, 'absorbed all the well-being of the Roman peasant.' When the Campagna of Rome was ruined by the Roman sewer, Rome exhausted Italy; and when she had put Italy into her Cloaca, she poured Sicily in, then Sardinia, then Africa. The sewer of Rome engulfed the world. This Cloaca offered its maw to the city and to the globe. Eternal City! unfathomable sewer! In these things, as

well as in some others, Rome sets the example. This example
Paris follows with all the stupidity peculiar to cities of genius."

Liebig very earnestly called attention to his views of the eco-
nomic bearing of the present system of sewers for large cities ; and
Victor Hugo stated the case very emphatically. These opinions,
though susceptible of some criticism, must in the main be received
as correct by nearly every thoughtful adult person of average in-
telligence. But questions of mere economy do not belong to our
present subject except as prosperity is promotive of health.

(28) In a previous chapter (§ 26) we studied the Mahopac disaster caused by draining a laundry and some water-closets into a
cistern ; the sewers are causing similar disasters every day. A
sewer carries the drainage of numerous laundries, numerous water-
closets and stables, manufactories and slaughter-houses ; the contents of the sewer after reaching the river are pumped up into
reservoirs for domestic use ; this dirty water distributed to every
dwelling-house, causes death and destruction. This unpleasant
subject is ignored as much as possible ; perhaps it may be thought
immodest to allude to it ; but to ignore it is death. The chemists
are analyzing the dirty water ; the microscopists are looking at it ;
and they are doing most important good work ; but more progress
must be made in these studies before we can introduce them into
our "easy lessons." Boston has her Cochituate Lake and Pegan
brook, and there is no actual need for chemistry to tell us that the
brook is dirty. New York has her Croton River and lakes, and
there is no need of microscopy to tell us that the hotels, the sani-
taria, and the manufactories discharge dirt into the reservoirs.

And Philadelphia has her Schuylkill River and her Fairmount
Park, where it is only necessary to walk about with the eyes open
in order to see that sewage enters the river at Manayunk. The
Delaware likewise receives sewage at Trenton, Bristol, Burlington,
and Holmesburg, as well as at Frankford and Kensington. The
pollution of rivers is the great difficulty with sewers ; all other in-
conveniences the engineers are rapidly removing.

(29) Many persons seem to think that the principal use of
sewers is to carry off excremental dirt; but this is really the
reversal of the order of things. Before the cities there were
streams of water, which were bridged for roads and streets ; the
bridges were gradually extended to cover over the streams till they
became continuous culverts; these necessarily carried the natural

contents of the streams, together with anything solid or fluid that might be thrown in.

The sewers of Rome, the Cloacæ, were so large that a load of hay could pass through them—*vehes fœni largæ onusta.*—LIVY. The city was supported in the air and navigated beneath—*urbe pensile subternavigata.*—PLINY. These great arches were never built merely to carry dirt into the river. In London it was formerly a penal offence to discharge excremental dirt into the sewers, which meant of course into the Thames, from which the whole city received its water; but the police could not prevent it, and it was eventually legalized, so that now all the houses are supposed to have soil pipes and drains emptying into the sewers. This involves an enormous expense for conduits to carry the sewage some miles towards the sea, before it is allowed to mingle with the water of the river.

In Paris the river seems to be kept reasonably clean. Attached to the houses there are tanks, made water-tight with Roman cement, and emptied periodically; the *voitures* dump their loads into barges, which carry the contents to poudrette factories. The comparatively clean contents of the sewers flow in separate conduits to irrigate many gardens, so that the water filters some distance through the soil before it can enter the general channel of the river.

(30) In the United States we are stupidly imitating London and Paris; and we follow with such tottering steps, that while they are trying to prevent the pollution of their rivers, we are trying how much dirt we can manage to throw in. We seem not to care to dispose of it in any other way. The sewers are written about as if they had been designed exclusively for the transportation of excrement to the river. It has even been proposed in official reports that storm-water should be carried off in some other way, and that the drainage of the soil should be excluded by making them water-tight. Of course those who write thus have not thought much about the matter. It is impossible with any known material to make a sewer really water-tight, as the settling on the irregular surface cracks any rigid material. The very most important function of the sewer is to drain the soil so as to render cellars partly under ground possible. Whole squares of land in the suburbs remain unoccupied, waiting for sewers to be built near enough and deep enough to drain the foundations. Our laws re-

quire the sewers to be placed deep enough to drain the soil four or five feet deeper than any ordinary cellar floor. The size of the sewers is especially calculated to carry off the storm-water. "They are designed to carry off one inch of rain .... in two hours, ... half a cubic foot per acre per second of the area drained," 1800 cubic feet per hour. "The sewage proper is 25 cubic feet per acre per hour." It follows that our sewers have 72 times the capacity required for the "sewage proper." Let us not think of excluding storm-water from the present sewers; and if separate

Fig. 24.

Fig 25.

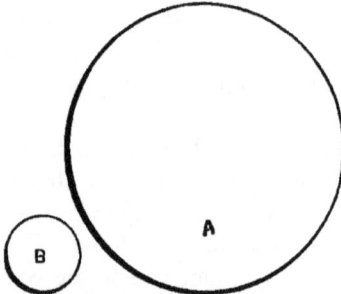

The Relative Size of Sewers. (A) to carry the water of rain-storms. (B) to carry sewage proper.

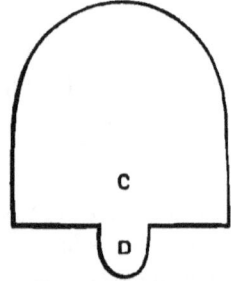

The Form of the Paris Sewers. (C) to carry the water of rain-storms. (D) to carry the ordinary drainage.

channels were a necessity in order to stop this pollution of rivers, we may construct new sewers for the "sewage proper." They may be very much smaller, of iron perhaps as a material better suited than earthen pipes or bricks.

(31) The sewers calculated to remove storm-water are deeply situated, and commonly drain the soil so effectually that the importance of this soil drainage seems hardly to be thought of; but there are situations where this action of the sewers needs to be supplemented. In grading for streets there is sometimes made-ground; it is generally low ground, and the sewers are not always low enough to drain it properly; hence houses on made-ground are likely to be unhealthy. Now, the hollows thus filled up are not always in proper situations for sewers, and they may be quite neglected in planning improvements. In the lowest part of these hollows there is sometimes a small stream; it has been proposed to construct a rough arch of boulders or any convenient rough stone, very strong and covered up no matter how deep, so as to act as a

deep drain; thus it is expected to keep the made-land in good condition perpetually without further care. This drain is supposed
to be strong enough to bear the weight of
any buildings that may be placed over it.
With this drain perhaps the houses on
made-ground would be as healthy as on
any other ground. Certainly there is so
much benefit to be expected that any cheap
material at hand should be arranged with
this object in view.

Fig. 26.

A Deep Drain of Boulders.
(Sanitarian.)

(32) The form of the sewers is a matter
of importance. Whether the stercoral pol-
lution of streams is to continue and increase or not, these drains
must be continued. They must carry off the water of rainstorms;
they must carry the continuous small stream of soil drainage; they
must carry the water of the rivulets that flowed over the surface
before the city existed; they must carry any additional waste water
from no matter what source. Hence, the sewer must accommodate
a continuous small stream, and it must be so arranged as to carry
an occasional flood. A common form is the *simple arch* across the
stream, a mere extension of the bridge. This for a sewer carrying
a continuous large stream, large enough to keep the floor covered,

Fig. 27.

The Common Sewer,—a simple arch.

Fig. 28.

Parisian Sewer,—a great
improvement of the
simple arch.

may be well enough, but for the ordinary sewer it is the very
worst. The dirt accumulates in this kind of sewer, and a small
stream of dirty water meanders on top of the slime.

A better form, perhaps the best, is that which prevails in the
city of Paris. This has a floor of masonry with a deep channel
for ordinary drainage, and there is the arch over it of sufficient

4

capacity for the storm-water. The ordinary flow is confined to its small channel; the rest of the internal surface is dry and clean till the rainstorm comes to give the whole arrangement a good washing. For large sewers this is certainly a desirable arrangement, as it affords a dry surface on which workmen may walk when entering for repairs, and the old single arch may be greatly improved by having its floor built up into this form. Perhaps even telegraph wires and gaspipes might be carried all over the city inside of these sewers, and thus be accessible for repairs without disturbing street pavement.

The best form for smaller sewers is the oval, narrow below and broad above. This has a narrow space below for the ordinary flow,

Fig. 29.

The Oval Form of Sewer.

without allowing much surface for the adhesion of dirt, and it has space above for the flushing when the flood comes.

The circular form is bad, as it affords more space for ordinary accumulations of dirt, so that it is more likely to be half filled with rags and obstructed in time of floods. The ellipse flattened from above is much worse, and the oval, wrong end upward, not quite so bad.

For the smallest sewers glazed earthenware pipes of the circular form are best, on account of the smoothness of interior surface, and the facility of placing them.

It is probably best that the floor and the lower part of the sewer, the part occupied by the ordinary flow, should be made as nearly water-tight as possible, as tight as stones and hydraulic cement can make it; the upper part, too, should be laid with hydraulic cement, for common mortar of lime dissolves out in many situations. The porous bricks of the upper part, with unavoidable defects of workmanship, cracks from settling, and openings for small inlets afford sufficient inlets for the drainage of the soil.

(33) The word *sewerage* is generally used to indicate the sewers themselves, the aggregate of bricks, stones, and mortar as just described. *Sewage* is the dirty water that passes through. *Sewage proper*, or rather sewage improper, is the mixture of dirty water and excrement, which through various domestic conveniences is made so liquid that it is floated off more cheaply than it could be carried in carts. As these conveniences are often death-dealing nuisances, we cannot avoid some description of them.

On entering an ordinary city house after a short walk in the open air, we mostly perceive a disagreeable smell that we call closeness; it is quite indescribable, a weak mixture of many smells; it comes partly from cookery, partly from the dressing of new carpets; it comes from kitchen-sink and water-closet; from slop-sink and bathroom; from laundry, from cellar, from sewers. It is sometimes called sewer-gas; it varies in smell and in chemical properties; it is simply foul air, dirty air. It may come from the sewer if any of the traps are defective; it much more commonly comes from decaying dirt in some part of the house.

(34) The *kitchen-sink* is faultless in mechanism; it is usually a simple cast-iron basin with a trapped outlet to prevent the passage of air from the sewer. But nothing seems too dirty or too clean to be washed in the kitchen-sink; at one time it is heaped full of dinner dishes, at another it is used to wash baby clothes. This is not as it should be; the fixed wash-tub is no remedy; the disgusting dirtiness might be mitigated by making the sink inconveniently high or of inconvenient form; but there is no remedy until people can be taught the constant habit of cleanliness in all things.

(35) The *slop-closet* has generally a cover of wood large enough to hold a bucket and to catch a little of the dripping that may be

Fig. 30.

The Slop Closet. The water instead of entering the funnel partly drips about the floor and soaks into the wall.

readily wiped up when thus visible. This is generally arranged by the plumbers and builders in such a way as to be a great nuisance: the wooden cover is made flat on both sides, so that when it gets wet the slops on the lower side drip about without entering the funnel at all; inclosed in a box this drip runs about the floor of the room, and even down the outside of the drain-pipe into the wall. This may be remedied by removing the box altogether, so that the drip may be seen on the floor and wiped up, or the cover may be changed somewhat to the form of a short funnel by tacking a string of caoutchoue around the lower side of the opening.

Dr. Andrew Furgus, of Edinburgh, has demonstrated that ammonia and other gases do pass through traps such as are ordinarily used; not only these gases but all kinds of dirty air pass and are diffused much more rapidly where there is no trap to be passed. The smell of these things does not come altogether through traps from the sewer, it comes more commonly from dirt or a slop of dirty water concealed about the sink.

(36) The *water-closet* contains all the nuisances of the slop-closet and some more. Besides the simple funnel that might empty directly into the trap, there is generally another compartment

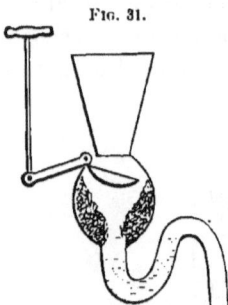

Fig. 31.

The Water Closet. The dirty arrangement.

partly concealed by a movable screen, a pretended valve, to hide the retained dirt. It would be hard to imagine anything worse than this; and the attempt is made to deprive it of offensive smell by such a lavish waste of water through leaky valves intentionally kept out of order, that no ordinary river is big enough for the city water-supply. We talk of wasting water by pavement washing; but this running of water through the wash-pave hydrant by some families for about ten minutes a week, is but a drop in a bucket as compared with the universal waste through water-closets by almost every family, for 10,080 minutes every week, throughout the summer. But all this waste does not remove the smell, for the shape of the receptacle is such that the adherent mass of dirt is only kept moist enough to make it decay the faster. It is not from the sewer that the offensive smell comes, the sewer-gas, the typhoid fever, the yellow fever, the cholera, the dysentery.

Perhaps the best arrangement in common use for even the interior of houses is the *common hopper* as used in the back yard; it is a simple conical vessel attached to the **ⵚ** trap, so that there is scarcely any surface above water except white porcelain; it cannot

be dirty much of the time among people with the ordinary in-
stincts of cleanliness. It may have ventilation flues downward

Fig. 32.

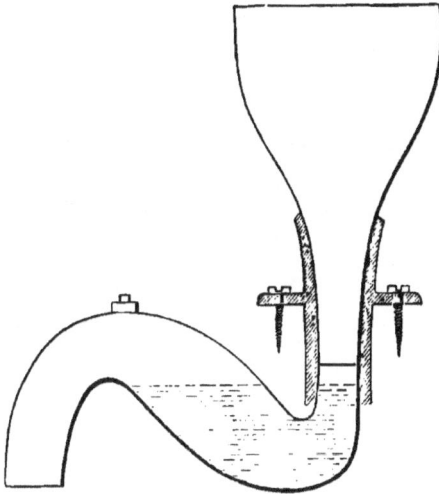

The Hopper Closet.

through the box, downward through the funnel, and from the discharge side of the trap ; but such flues should be kept separate till

Fig. 33.

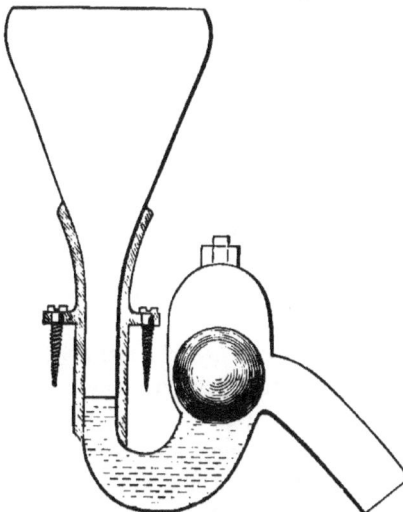

Hopper Closet with Check-valve.

they discharge into the open air, for otherwise the foul air may go
up one flue to come down into the room by another. There is no

real necessity for a wooden box to conceal the arrangements, for the seat may be suported like a chair on suitable feet.

If offensive gases pass through these traps, as the experiments of Dr. Fergus demonstrate that they may, or if typhoid germs, like so many minute frogs, hop into the water of the trap' on one side to jump out at the other side, nothing can be more effective than a check-valve, the sewer-gas check-valve as we see it advertised among plumber's supplies: this is nearly perfect and only wants enlarging a little to suit traps of different sizes. For durability it has been suggested that the valve should be made of hard mate-

Fig. 34.

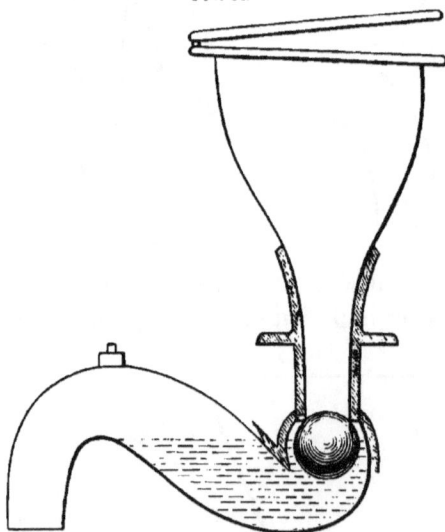

Hopper Closet with Hard-Rubber Floating-valve.

rial, such as marble or glass, and its socket of brass; so that the wear of use may tend to perfect and maintain the exact form, but a good ball of gutta-percha in this situation would answer the purpose and last a long time.

(37) *The Ship's Water-closet.*—If anything better than this is called for, we would recommend the arrangement common on shipboard for use below the water-line. It may be placed even in the hold, and only requires the raising and lowering of a pump handle to empty the basin and completely wash it out. If used in a dwelling-house, the valves may be merely check-valves of suitable size, so that if the water fixtures leak, the surplus may flow through without material damage. This machine, with its two valves and

its plunger, its stuffing-box and its pump-handle, may be rather expensive; but it can hardly cost as much as the lives of people are commonly worth.

Fig. 35.

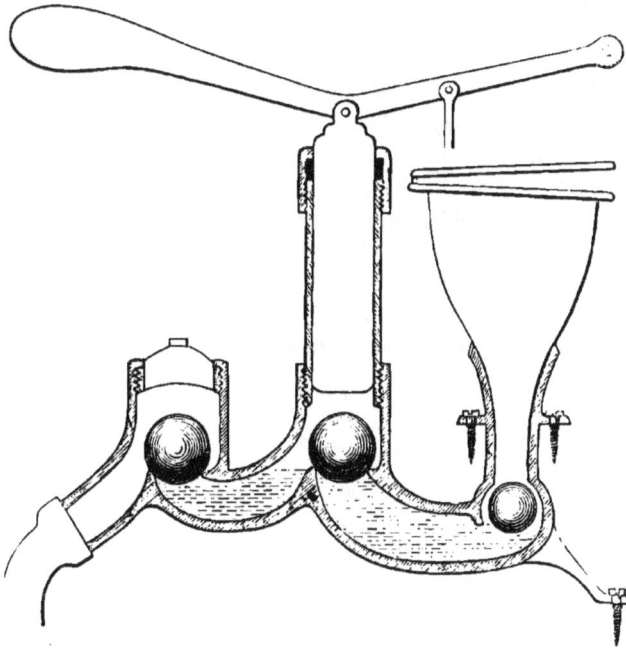

The Ship's Water-closet with Ball-valves. Any number of valves may be added and a ventilator to each trap, and the floating-valve may be omitted.

(38) *The house-drains* are expected to carry off all the dirty water and fluid waste from the house.   When of small size, from wash-basins, bath-tubs, and kitchen-sink, they are best made of lead pipe; and when larger they are of cast iron with leaden joints. Terra-cotta pipe is used to connect these drains with the sewer, but should not be allowed inside the walls of the house.   These drains should never enter the cellar floor, but should be carried by the shortest possible course beyond the walls.   If a drain must cross a cellar, it is best to arrange a shelf for it on a wall where it is constantly in sight.   Even royal perfection of plumbing cannot make these things quite safe in a dwelling-house, for Prince Albert lost his life by a defective house-drain under his library, and his son, the Prince of Wales, suffered a serious illness and a narrow escape with life by a similar incident.   Quite recently we see a material change in these arrangements; formerly the drainage from the

kitchen went by a pipe through the cellar, in the cellar floor, from rear to front, and so on to the sewer, or near the outer wall there was a culvert leading to the sewer. The culvert was so large that it might be entered for cleaning, two feet wide by four feet high, but it was never cleaned; the filth accumulated from the dirty water till the culvert was full, and then the water kept a little channel for itself on top of the slime, and for that matter these streams are still flowing through their little channels in the accumulated slime of these culverts, that have not been cleaned out for a hundred years. This is one cause of the great mortality in some parts of the older cities. This is all changed, or changing for the better. Well-jointed pipes of lead and iron pass directly through the rear wall into a small garden, where they are united with pipes of glazed earthenware; these are continued across the rear of the lots, at sufficient depth and with sufficient fall, to the

FIG. 36.

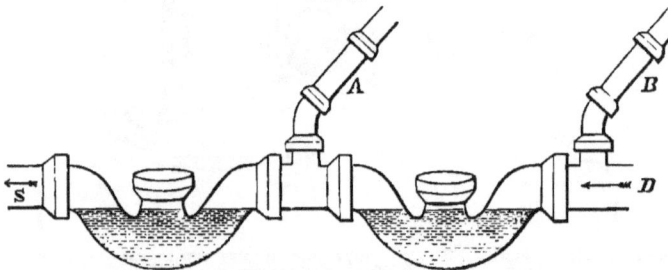

Ventilated Traps. *A, B,* ventilators; *D,* drain; *S,* sewer end of the drain.

public sewer. The building laws of our cities require the reservation of sufficient land for these small gardens, and the builders generally favor them; but old buildings are not easily changed to conform to the plan, especially as it requires a right of way for a drain across a neighbor's land. It seems difficult to carry out this improvement in small houses, as it is perhaps impossible to prevent the people from forcing rags, broken glass, and other improper things into the drains. When the drain is through their own cellar they sometimes learn to do better. Whether these drains should be trapped or not is still a question under discussion. The traps themselves are pools of dirt, but if there is any necessity for them, there is no difficulty in having any number and all ventilated, as in the diagram (Fig. 36). The principal use of the traps is to catch rags and broken bottles before they get too far from the persons causing the nuisance.

(39) *The ventilation of sewers* and drains is a matter of importance. The large drains and sewers cannot be made air-tight, but the metal pipes inside the house may be made water-tight by good mechanical work. The air of any ordinary room with doors and windows closed, notwithstanding the porous material of the walls, becomes unpleasant in the course of a few days, smells musty, and is unfit to live in till the room is opened for ventilation, and perhaps warmed. Drains and sewers, besides the confined air, contain dirty water, decaying and giving off a much more offensive smell. Now, this dirty air may be diluted by constant ventilation, or it may be concentrated by constant shutting up, so as to cause sudden death if incautiously or accidentally entered. We really have no means of getting rid of this offensive air except by infinitesimal dilution, and this we are apt to call "getting rid of it altogether." The natural movements of ventilation, as produced by changes of temperature, must be borne in mind. Warm air is specifically lighter than cool air, mere foulness affecting the weight so slightly that, as influencing movement, it need not be considered at all; hence, when the sewer-air is cooler, as in a summer day, it tends to remain in the sewer or to flow out at the lower end; but when the sewer-air is warmer, as in winter, it ascends by every possible opening, being displaced by the heavier cold atmosphere. This foul air does not mingle instantly with the cooler pure atmosphere, but in obedience to the laws of gravitation it ascends to the higher regions of air, quite out of harm's way. Apertures larger than the pores of bricks are desirable, but where or how many we are not prepared to say. A separate ventilation-flue in every chimney-stack is a good thing for houses, but certainly sewer-air of the public sewers should be carefully kept out of it. Openings for other purposes act as ventilators, such as inlets, outlets, and entrance openings for repairs. It has been suggested that manufactories using much fuel might draw air for the furnaces directly from a main sewer, but the air would enter by the nearest opening, so that the effect would be small unless a large furnace could be used near the outlet in such a way as to capture the foul air in the last section of the sewer just before its escape. Such, then, is perhaps the present limit of our power in this matter. We may make as many man-holes as we please along the central line of the street for the escape of sewer-air in cool weather, and we may have a furnace near the outlets to capture and burn up the dirty air just before its escape in warm weather.

(10) The stercoral pollution of rivers, besides the disgust, is a very serious cause of disease and death, and perhaps this subject may be usefully illustrated by a brief reference to the natural history of some of the entozoa, such as the tapeworm. The *tæniadæ*, tapeworms, vary much in size and form, and about ten species have been found infesting the human body. The species in nature are very numerous and infest nearly all vertebrate animals; they are perhaps not so numerous as the butterflies, but they are developed through about the same number of similar changes in form. The eggs of the butterfly hatch, and we have caterpillars crawling about, doing mischief, and changing coats (moulting) till they get old enough to wind themselves up in cocoons of silk or to hide away in a corner. They are now called *pupæ* or *chrysalids*, till their time comes to show themselves as full-grown, full-fledged butterflies. Similarly, (*a*) the *ova*, the eggs of the tapeworm, when received into the stomach quickly hatch, and (*b*) the young, called *cysticercus* (bladder-tail), penetrates to nearly all parts of the body; he is made up of head and tail, the tail being a bladder full of water, his head is nearly hemispherical, but he has no mouth, and his teeth, if he has any, are on the top of his head. His tail must change much in form and even be drawn out into a string sometimes, so as to get it through the small hole made by his gimlet of a head. He probably starts on his travels without the bladder, a mere string pointed at the ends, and penetrating a bloodvessel he is carried to the liver or to some other part of the body, accordingly as he first falls in with a branch of the *vena portæ* or with some other vein, a threadworm in the blood (a *filaria hæmatobia?*). The *cysticercus* may cause death by penetrating the heart, the eye, the brain, or the spinal cord; he gets too old to travel, and he stops for a rest; he turns his neck inside out, and his head settles back into the middle of his bladder. If he finds himself too near a joint so as to be much disturbed and pinched, he stretches out his head and starts off again till he finds a more suitable place, and then he remains a little bladder of water with a dimple on one side. He has now reached (*c*) the third stage of his existence, corresponding to the chrysalis of insects; and as caterpillars do not all make cocoons, so the cysticercus at rest assumes various forms and receives various names; they may be called hydatids as a general name. The *hydatis cellulosæ* in pigs is called measles; the *hydatis mediocanellatæ* (= *echinococcus*) has been found in the human liver, and has been produced in young cattle and in some other animals by mixing the ova with their food;

the *hydatis marginatæ* (= *cœnurus cerebralis, braxy*) has been found in the brains of sheep, causing staggers.　Even the same species varies much in form : the ova of *tænia serrata* develop into *cysticercus pisiformis* in puppies, into *cœnurus cerebralis* in lambs, and into something very like *ascarids* in the liver of rabbits.　Cruel Dr. Leuckhart, to coax these lambs and rabbits to eat bits of tapeworm and then to knock the poor things on the head, and cut them into such little pieces merely to see what had become of the hateful mess!　The most dangerous form is the *hydatid tumor.*　The echinococci of several species—tenuicollis, marginatæ, pisiformis—possess the faculty of *parthenogenesis,* and thus multiply within their thickened sac till they constitute large tumors, even as large as a child's head.　The next event in the history of these creatures is to be eaten up, when they quickly develop into (*d*) the regular tapeworm ; and with the teeth or hooks on top of his head he just

FIG. 37.　　　　FIG. 38.　　　　FIG. 39.

The cysticercus travelling.　　Cysticercus at rest.　　Cysticercus mediocanellatæ.

hangs himself up.　He does not even swallow his food, for the surface of his body does duty as stomach ; he just hangs and drops eggs for new generations of (*filaria hæmatobia ?*) cysticercus, hydatids, and tapeworms.　Many other things, however, may be in the blood and be mistaken for filaria.　The young of the *filaria medinensis* must be carried in the blood ; the *trichina spiralis* reaches nearly all parts of the body by the current of blood ; nearly all the *entozoa* infesting the liver must be carried there from the intestines by the vena portæ ; besides, there are such vegetable organisms as the threadlike bacteria (*bacillus, anthracis,* etc.), which besides causing other deadly diseases, appear to block up one minute artery at a time, and thus one finger or toe after another, nearly all in succession, are destroyed in that awful condition called *lepra anæsthetica,* till fortunately a knot of them plug up a large artery and end the misery.

(41) Among the more noted species of tapeworm may be mentioned the following :

Bothriocephalus latus, the broad tapeworm, tænia lata, t. vulgaris, t. grisea, t. membranacea, t. tenella, t. inermis, t. prima, t. acephala, lumbricus latus, halysis lata, dibothrium latum, comes first because he is the biggest and has most names. He is about an inch wide and wonderfully long. He is not common with us, but is occasionally found among new immigrants. He flourishes in the North of Europe. The cysticercus and hydatid of this species have not been recognized; and not having the hooklets on the head, necessary under the microscope for the certain recognition of echinococci, we should find them difficult to determine even if we had them in our hands.

Tænia solium, the solitary tapeworm, t. longa, t. cellulosæ, t. cucurbitina, t. secunda, t. articulos demittens, is perhaps the most common species in this country. Its hydatid in repose, hydatis cellulosæ, gives the peculiarities to measly pork. The stories told of the length of this creature are really wonderful, reminding us of the person who was so very sick that he threw up three black crows. Dunglison says : " It is said to have been met with 600 feet long."

Tænia mediocanellata, Leidy thinks, may be as common as the preceding. It infests principally the Abyssinians and others who eat raw beef. Hydatids of this species, found in a human liver, were fed to various animals, and the following were infected, mostly by this means: Ass, cat, and camel; cow, chamois, and deer; goat, giraffe, and horse; kangaroo, monkey, and pig; ox, sheep, squirrel, and zebra.—*Cobbold*.

(42) The fully developed tapeworm causes much pain and discomfort, but it does not seriously endanger life in this form, and it is so rare that to meet a patient suffering from it seems to afford a real gratification to the physician, for he knows that by a little judicious starving, alternated with an occasional dose of pumpkin-seeds and castor oil, this curious animal will soon be laid out on a board, to be put into a vial of glycerin and water for preservation. Gamgee thinks that fully three per cent. of all the pigs in Ireland are affected with hydatis cellulosæ, measly. Thudichum a few years ago reported that nearly every sheep slaughtered in London was affected with braxy, hydatis tenuicollis. Hjaltelen calculated that twenty per cent. of all the deaths in Iceland are from hydatid tumors (echinococcus marginatæ) and tapeworms; while others

have estimated the proportion at about fourteen or fifteen per cent. There are about six times as many dogs as men in Iceland. Every dog is affected with tapeworms, and the eggs soil the grass and water and nearly everything. The sheep must eat the eggs on the grass, and hence the universal braxy. The sheep are slaughtered and the braxy part thrown to the dogs, and the perpetuation of the race of tapeworms is thus provided for. The people probably do not eat braxy mutton without cooking, but they drink water without boiling, and some vegetables without cooking, and hence twenty per cent. of the deaths are by tapeworms and hydatids.

But what is to be done about this matter? Market inspectors should be able to recognize braxy mutton and measly pork; but the butchers may cut off the conspicuously diseased part and conceal it. In Iceland it has been proposed to treat all the dogs regularly with tapeworm medicine; but the dogs resist and the people are careless. Perhaps in time a habit might be inculcated of feeding the dogs exclusively on cooked food.

(43) The human intestines, and nearly all parts of the body, are infested by very many deadly animals besides tapeworms. Some of them, as trichina, strongylus, bilharzia, and filaria, are pretty well known; others, from their minuteness rather than their rarity, are seldom seen, and their manner of life is unknown. But, enough! is it really necessary, in an enlightened country, in the latter part of the nineteenth century, to pursue the natural history of these animals further, merely for argument to prove that stercoral matter is dangerously poisonous, and, if swallowed, capable of producing disease and death?

(44) There is an effort to stop cesspools from leaking into the cisterns; but this is done by emptying the cesspools and sewers into the rivers. We are told that organic matter (stercoral matter) thrown into the river is quickly diluted and oxidized, so that if the water flows a mile or two it becomes sufficiently pure. There is some truth and much fallacy about this statement. The ova and embryos of tapeworms and some other entozoa do perish and decay in the course of time, if cast into the river; but some of them are a long time before they perish, and some probably live in this situation long enough to become the parents of millions, and thus to continue indefinitely. Physicians cannot resist the accumulating evidence that typhoid fever infection is composed of definite organisms, "probably living." Dilution separates the infective germs, so that any particular cup of the water is less likely

to contain the deadly portion ; but the numerous cases of typhoid
fever prove that the poison is still present and in such condition as
to cause disease and death. The deaths from this disease, caused
in this manner, in each of the large cities, amounts to several hundreds every year.

(45) The problem for engineers is, How are the people of large
cities to be supplied with tolerably pure water? with water not
polluted by the drainage of filth from their own homes?

In London some wealthy people solve this difficulty for themselves by drinking only mineral-water—not nauseous doses of
Seidlitz or Saratoga, but the purest sparkling spring-water that
can be found. It is good table-water, good mineral-water for the
table, the queen of table-waters. There are large bottling establishments, where the water is duly corked, sealed, and labelled,
with every possible precaution to prevent deception. Sometimes
the water is charged with a little more gas, so that when the cork
is drawn it pops like champagne. Some of our apothecaries sell
these pure table-waters from across the ocean. This supply of
pure drinking-water to a few wealthy families by carting it around
in bottles and demijohns is possible, but it is not possible to supply
the people of a large city in this manner.

In some large cities, notably in London and in Paris, the attempt is made by means of large sewers parallel with the river to
intercept all the sewage and thus conduct it to a safe distance, and
even to utilize it as manure in the fields. To discharge the dirt thus
at a distance might do for a seaport, but for any interior city it
would be a gross outrage. It would be to recklessly pollute the
water-supply of the next city.

The utilization of sewage for manure, heretofore, has not been
profitable, because of the profuse wasting of water, which so dilutes
the material that *it has not enough money value to pay for the labor
of applying it to the fields.* If this diluting can be checked, by the
use of water-meters, for instance, perhaps manufactories of manure
and hydraulic cement, with facilities for irrigating gardens with
the comparatively clean water, may be made to pay—may be possible.

(46) Another device that seems plausible is in use at Lowell,
Mass. (*Report of State Board of Health*, 1874.) This is to filter
the water before it enters the reservoirs. A suitable location is
selected, a bank of sand or gravel, a mile or so wide, near the river
from which the water is to be drawn. A series of wells may be

made at sufficient distance from the river to insure perfect filtra-
tion, and of sufficient depth so that the water will flow as fast as
pumped out; or, instead of the series of wells, there may be one or
more elongated structures, *filtering galleries*, extensive enough to
filter all the water of the river if needed. The river must be very
bad if it cannot be made to supply good water in this way. At
Lowell the filtering gallery is but one hundred feet from the river,
and the water is abundantly supplied with the usual inorganic
salts of well-water, so that probably very little of it comes from the
river. But, as we have said, this is a question for the engineers.
We insist that water should be supplied without systematic pollution with sewage.

(47) Another kind of water pollution occasionally receives more
consideration than it is fairly entitled to. Various manufactories
discharge into the rivers waste material. Tanneries discharge large
quantities of exhausted bark, mixed with hair and filth of their
peltries, with some little dye-stuff. Exhausted bark is about as
innocent as any organic material can well be—no worse than sawdust. The filth and hair are worse, being offensive by their decay;
but the inorganic germs of disease can hardly be imagined to maintain their virulence through the operation of the tanners' vats.
Dyeing and bleaching establishments discharge quantities of colored
material. Indigo, with its frightful color, is about as innocent as
starch or flour. Logwood and brazilwood are about as poisonous
as oakwood, and their chips and shavings are about as bad as the
sawdust from the lumber mills. Turmeric, in the quantities present, cannot do much harm, as, alternating with annotto, it is used
to give color and flavor to oleomargarine and high-colored butter.
Such material should not be thrown into the rivers, especially
as farmers gladly haul it away to enrich their fields—even pay for
the privilege. Similarly of the mineral waste, iron and tin, lead
and copper, arsenic and bismuth, calcium and chlorine, chromium
and iodine, magnesium and manganese, mercury and phosphorus,
potassium and silver, sodium, sulphur, and zinc. These, in the
relatively small quantities present, with time and circumstances
favoring, unite in accordance with their chemical affinities. In part
they precipitate, and in part form about the same inorganic salts
that are always present in all *good spring-water*, and serve to dis-
tinguish it from distilled-water or rain-water. It has even been
objected to the use of water that does not contain those salts necessary for healthy nutrition. It is quite possible to conceive that a

manufactory of Paris green, or the like, might dangerously poison a stream of water; but the organic germs of disease, amœba and bacteria, trichina and filaria, bilharzia and strongylus, cysticercus, echinococcus, tapeworm, and the like, could hardly pass through the dyer's vats and live. The danger from manufactories is from the sewage of their villages.

CHAPTER IV.

ABOUT PLUMBING.

(48) SOMETHING about plumbing has been learned in the school of experience, and following illustrious examples it may be well to redintegrate a little and tell all about it.

Some years ago a naval medical board were sitting among books and papers discussing "What shall we do next?" when there arrived several telegraphic orders, all worded as follows: "You will report immediately to the commandant of the navy-yard for duty in reference to cholera." It was suggested that "We may as well shut up our books and put on our hats; but *while we are all here* let us sign up the record of proceedings." The commandant was glad to see us. "There is cholera on board the receiving ship. Suspicious cases were reported yesterday with one death, another death last night, and several undoubted cases. The surgeon of the ship is pretty tired, and has enough to do with the rest of the crew. About fourteen doctors are ordered by telegraph, but as yet only one has reported. On the recommendation of the surgeon, the yard tug has moored two other ships in order that the men may be removed and distributed, and has now gone for another to be used for a hospital ship; this will take nearly all night, and you can inspect her in the morning and report as to her fitness for the purpose. If this meets your views we can transfer the sick, and if one vessel is not enough we will have another as soon as possible."

"This seems to us the best possible arrangement. We will now go and see the sick on board, and we will be ready to occupy the hospital ship as soon as it can be properly moored." . . .

The next morning a dispute is heard in front of the commandant's office, near a bier sitting on the footway.

"But it is my husband," the woman persisted.

"I cannot help it; he is dead, died of the cholera, and I have orders to remove the body. I must start with it to the cemetery as soon as the other arrives, for which I am waiting."

"But I want to have a funeral."

· "If the undertaker were here I could deliver the body to him, but it must not remain here."

"But I want a little time to get money to bury him."

"If I take the body to the cemetery it will cost you nothing; everything is paid for, and you can follow in carriages with your friends and have your funeral services in the chapel at the cemetery." . . .

A number of carts arrive with strange-looking loads, and the head teamster advancing touches his hat to report: "Sir, I have been ordered to deliver some hospital furniture." There are four or five carts. The bedsteads are painted red, and have been screwed together in the night to save time. Each cart has some bedsteads, and the mattresses and blankets, sheets, pillows, and pillow-cases are packed in below.

"Very well, the furniture is to go aboard the hospital ship; go as near the end of the wharf as you can without getting in the way, and wait."

An officer arrives to report: "The captain of the tug sent me to report that the mooring of the hospital ship is nearly completed, and to ask if there are any further orders for him."

"Thank you; we will go on board with you."

We start for our inspection. The ship is in excellent condition; she has come in from a short cruise, and is as clean as scrubbing can make her; she has a donkey boiler and steam pipes to warm throughout, gun-deck space enough to accommodate twenty-five beds nicely.

"Well, captain, you have her secured?"

"Yes."

"No rattling of chains, then, for fear of dragging in case of a gale?"

"She has anchors enough to hold a line-of-battle ship."

"Thank you, captain, and now there are some bedsteads at the wharf; please bring them aboard and the hospital stores—everything that is in the carts." . . .

"Well, we are getting on bravely; the engineer is firing up and we shall soon have steam in these pipes nice and warm; and

see here, this extra boiler, with its furnace just in the right place
to burn up dirty bedding; the cooking galley in complete order;
nice rooms for the officers, all the better for being without carpets
or bedding."

" While we have the tug at our disposal let us bring the patients
aboard."

" Yes, everything is ready as soon as the beds are made up, and
if once this tug gets away from us she will be off elsewhere, and it
may take us a whole day to get another."

" I noticed yesterday that the captain of the receiving ship
thought the sick very well situated where they are now that the
crowd of men are away, and I feel that he will give us trouble
unless we go to him with a peremptory order."

" As I am the youngest, I will stay to fix up the beds and the
furniture while you hold the tug, and the order will do no harm
even if we should not need it." . . .

" Commodore, we are ready for the patients aboard the hospital
ship, and would like to have your permission to make the transfer."

" Is the vessel sufficiently warmed? Captain —— has been
here and suggests that the receiving ship is right comfortable, and
to remove men exhausted by sickness to a vessel that has not been
occupied for months before it is well warmed would be dangerous;
it might be fatal to some of them."

" Certainly, but to move these men quietly takes time. The
fire is lighted, steam is up, the vessel is warming, and before the
men can be moved she will be warm. We think it important to
make the transfer during the warm part of the day."

" Well, then, transfer them as soon as you conveniently can." . . .

" The beds are ready, the vessel is warm, but our fireman reports
some small leaks about the pipes. He has shut off steam and has
gone to the plumber's shop to see about some little repairs that he
thinks may take an hour."

" I go for the transfer at once, even if we cannot use the steam-
heaters for a week."

" Probably it will not be so bad, but even if we had no steam-
heater at all, I presume there would be no want of unanimity on
this question." . . . . .

" Captain, we are now ready to go with you to the receiving
ship; we expect to bring the sick aboard on the deck of your tug."

" Aye, aye, sir." . . . . .

"Captain, I have come to relieve you of the care of the sick,— to transfer them to the hospital ship."

" But when I saw the commandant this morning he agreed with me that the ship could not be properly warmed for sick men in one day, and that they had better remain here till the vessel could be made comfortable for them."

" I have just been to the commandant's office to report the hospital ship furnished, warmed, and all ready; and he has given me the order to make the transfer at once, so as not to lose the warm part of the day." . . . . .

" The officers have all gone to the new receiving ships, so that the transfer will not be very regular."

" We will not be particular about muster-rolls; there are enough willing, strong hands among the nurses; the assistant surgeon ordered for duty with the sick we will take with us." . . . . .

" All ready."

" Now, men, we are going to take you to the hospital ship that we have been getting ready for you. No, my good fellow, do not look about or think of getting up; there are plenty of men here to carry you; shut your eyes if you can; do not be afraid, and do not lift a finger. Here, take him up quietly, bed and all, that is right." . . . . .

"Come this way; plenty of time; steady now."

" Here; lay him quietly; just here. Lie still and let me fix you a little. There, that will do—and I will put this over your face to screen your eyes a little. Here now, put him right alongside of this one—that will do. Another one here, so as to make a line right across the deck." . . . . .

" The sick are all aboard."

" Their clothes?"

" The nurses have gone for them."

" Bring them along, nurses and all." . . . . .

In three minutes we were alongside the hospital ship, and a gang of frightened plumbers left in some haste from the opposite side. The sick were soon nicely in their beds, apparently improved by the trip.

" We have done a good day's work, and as we have no further use for the tug I will just give the captain our unanimous vote of thanks, and let him depart for other fields and victories. I shall bargain for him when we take the hulk to the quarantine station for disinfection." . . . . .

Our fireman reported that: "The plumbers took the pipes apart, and in their hurry carried away one piece. They wanted to put in new pipe, so that they would not have been ready to warm the vessel for a week."

"What can be done?"

"I have put the pipes in place, and if I could only find the missing piece there would be no trouble. The pipe was arranged for the steam to circulate so that the condensed water returned to the boiler. The loss of one piece breaks the circuit, but it may not take long to fit another piece of pipe."

"You must do the best you can while I go ashore and see what can be done." . . . . .

We soon found the chief engineer of the yard. "Trouble?" "Yes."

"I saw the plumbers running as if there had been a mad bull after them; and I sent a foreman to see about it. He reports that there really was nothing the matter at first, only some little leaks, that the fireman would have repaired by painting the cracks and wrapping with twine; but, unfortunately, he went to the plumber's shop for the paint and ball of twine. The foreman who went aboard with him wanted to make a good piece of work, and so set a gang of men to take things to pieces. I have about a dozen men hunting for the lost piece of pipe, and two men making a new piece. This is slow work, as we have delays for want of a boat to carry them aboard for measuring, etc."

"Are not your men afraid, the same as the plumbers?"

"No; they are firemen and coal-heavers of the navy; and you know they will not flinch when ordered on duty, especially as there is a medical officer aboard to notice and encourage them."

"I thank you in advance for carrying us through this trouble, and I hope you will succeed in establishing the circuit of steam-pipes before the ship gets chilly." . . . . .

The next morning was bright and pleasant. "We did very well during the night, but towards morning it was a little chilly. About sunrise the ship became nice and warm. We had one death this morning." . . . . .

The rest of the history of the hospital ship is as dull as the history of a country without any wars. There were four deaths after the transfer. Some new cases of cholera and diarrhœa came from the receiving ship, but all recovered. We towed the old receiving ship to the quarantine station. We treated her to two large carboys of crude carbolic acid, and three barrels of iron sulphate (copperas), enough as we judged to endanger the lives of any amœba or bacteria that might be aboard; and there we left her. With our present views we would have treated her to the fumes of a few pounds of burning sulphur. Thus ended the cholera epidemic.

(49) Hence we know that *steam-pipes, gas-pipes, and water-pipes, all kinds of leaky pipes of moderate size, may be instantly repaired by giving them a coating of paint and a wrapping of twine; and if by any chance a small piece of pipe is missing, a piece of gum-hose may be cut with a jack-knife to connect the open ends of remaining pipe so as to complete the circuit.* If we had possessed this "little knowledge" a few hours sooner, it would have prevented some suffering and perhaps it might have saved one or two lives.

(50) Our second lesson in plumbing was not so interesting; there was nothing tragical about it The water-supply of the Naval

Fig. 10.

The Wipe-joint.

Hospital . . . is rain-water collected in large tanks near the roof; the overflow is received into cisterns, from which it is pumped up in case of need in dry weather. The engineer is a tinsmith who has had the care of a small lathe and a planing machine at the navy-yard; he was recommended by the chief-engineer of the yard as a good jack-of-all-trades and just the man for the place; and having been instructed that he would be expected to do all the small repairs himself, not many days elapsed when he produced specimens of his plumbing work; he was evidently proud of his wipe-joints.

Dry weather brought trouble. " I am afraid I have an enemy somewhere."

" Why so?"

" Last evening I pumped up 3000 gallons of water and the tank was empty this morning; no water used in the night."

" Well, pump up half as much to night; keep your eyes and ears open." . . .

" I have found out about losing water from the tanks. . . . I went tip-toeing around after midnight. I shut off the water, and

started it in one pipe after another till I thought I had traced a
leak in the vacant ward of the south wing; at last I heard a buzz
like a big fly, in a water-closet occasionally used by officers. It
was a very small leak, but there is no loss of water since I repaired
the valve."

Hence we know that *a very small leak in a water-closet, such as
no plumber discovers in a common inspection, may waste as much
water as a hundred families can actually use.*

(51) The third lesson was little more than a repetition of the
second. A naval hospital . . . . is supplied by the city water-works
through a meter that says "click-click" as each gallon enters the
building; the engineer is a capital blacksmith and does all small
repairs in his line, but the plumbing and gasfitting were done by
regular workmen called in as required. It is easy to see that more
time was spent in coming and going than in doing the work;
thus a change became necessary in the interest of the hospital fund
as well as for economy and morality in general. An expensive
set of tools was purchased, tube-cutters, taps and dies, and
wrenches. These had to be tried; joints were unscrewed, cleaned,
oiled, and screwed up again; leaky valves were found and repaired or new ones substituted. The steam-pipes and the engine,
the gasfittings and the plumbing were greatly improved in appearance, and the saving of money was quite considerable, as the expense was now reduced to little more than the interest on the
moderate sum invested in tools.

The end of the quarter came around when the bill for water
should have been presented at the extravagant rate of three cents
per 100 gallons; the bill did not come; the hospital purveyor
called for it, and they promised to send it "to-morrow;" he called
again and again; there was "something the matter," the meter did
not register properly, and they did not know how much water to
charge for. The meter, however, had been saying "click-click"
quite punctually, but it had registered less than half the usual
quantity of water.

Hence we infer with a degree of probability that comes little
short of demonstration that, *about two-thirds of all the water pretended to be used in our cities, is mere waste through defective plumbing, so concealed as not to be found in the ordinary inspections.*

In order that housekeepers may know how much water they are
wasting, all the water *except that of out-of-doors hydrants* should
pass through water-meters.

CHAPTER V.

(52) A worker in tin and sheet-iron had his home and his shop on the shady side of the street, and he did excellent work in fitting furnaces and ranges for dwellings in the neighborhood. The retail dry-goods shops crowded him and rents went up—he must move.

"I hope you will not move far, for I want you to work for me sometimes." He would not move if he could help it; and so we became owners of a margin in real estate, with the understanding that he should be the owner as soon as he should be able to repay the margin. Inspected from the opposite side of the street, the brick front stood reasonably well, but, on coming into possession, we found a sorry nest of rats and rottenness.

The rats were driven away by filling their holes with sand and a few handfuls of copperas; there was a general whitewashing and papering, and my man of "furnaces and ranges" felt comfortable. Winter came, and the cellar became the play-room of the children, and they were unhealthy. Spring came, and the alley being obstructed with ice, the rain and melting snow flooded the yards, and made some of the cellars wet. It was time to call the attention of the Board of Health to the nuisance. The privy well was full, as the inspector reported, but nothing else was found requiring the attention of the Board. We discovered that there is no sewage-drain in our cellar; the wells of this row are so arranged that one well, with two closets over it, serves for two houses, and so there is a drain in each alternate cellar (Fig. 41). But through the rat-holes, the air of all the cellars was about equally bad.

The Board has no authority to build a sewer, no money for hardly anything; next year perhaps something may be done. After waiting a year, the agent reported the same things again, with the addition that the tenants complain of offensive vapors from the dye-house.

(53) This was answered by a resolution of the Board, "That the privy well on premises 1302 . . . by reason of its proximity to premises 1300 . . . is a nuisance . . . The Health Officer be directed to abate the same by having said well cleaned . . . filling with clean earth, and constructing in lieu thereof a hopper closet, with . . . proper sewer connection—work to be done in accordance with law and the rules of the Board." "Rules of the Board," as interpreted by inspectors and workmen, meant that we were expected to build a sewage-drain through the length of the cellar;

we made a remonstrance in the shape of an appeal to the Board ;
were invited to meet a committee ; this meeting happened when
members in twos and threes were discussing the recent arrival of a
vessel under exciting circumstances, and when poor women were
waiting to complain that employés of the Board had charged them
too much for whitewashing, "the whitewash was not white, and
the work was not well done any way:" there was no chance for
reasonable conference. The appeal was dismissed ; but the agent

FIG. 41.

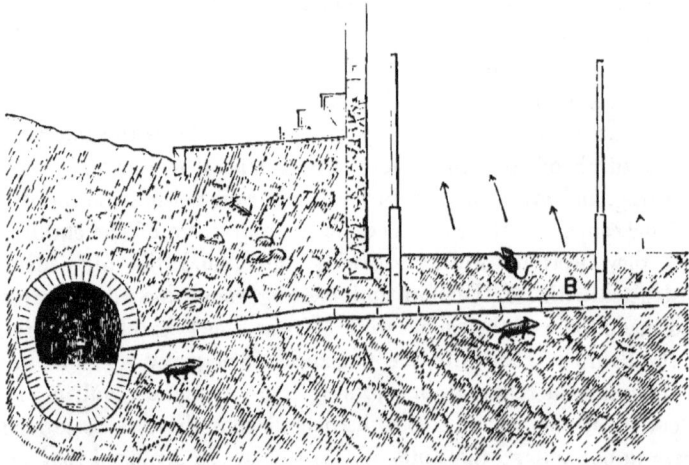

The sewage-drain in the cellar—a nest of rats and rottenness.

received a hint, that as we started the matter ourselves, probably we
need not do the work just now, unless we chose.

(54) We did not want the affair to die in this quiet way, and so
tried a second appeal. We recited some incidents of the conference : One gentleman thinks the only difficulty is the length of
drain ; our objection to the thing is that we are ordered to place a
sewage-drain in the cellar—to bury it—to hide it so that no man
can see it. Another thinks each house should have its own drain
to the sewer : with reasonable exceptions, this is certainly a good
rule ; but the drain of filth through the cellar should not be of
earthen pipes buried out of sight.

In conclusion, having been ordered by "Resolution of the Board,"
as interpreted by the committee, and by the official inspectors, to
place a sewage-drain in a cellar and to hide it there, we objected to
do any such thing. We proposed to arrange the matter by using
iron soil-pipe caulked with lead, and attached to the cellar wall so
as to make connection with the sewer in the street north ; we hoped
that this would be accepted as the performance of the order.

A few days brought the answer that we might do it in our own way, and that the Board would approve of the use of the iron pipe as suggested.

(55) The work was done : the waste at the hydrant enters by a bell-trap; other fixtures have ∽ traps; and the iron pipe, about 70 feet, has a good slope; in the cellar is attached to the wall,

Fig. 42.

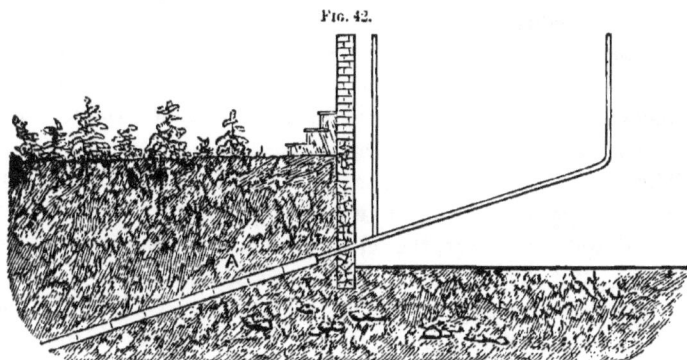

The iron soil-pipe attached to the cellar wall, may drain through the garden, or may connect directly with the sewer.

penetrates the front wall near the floor, and so it proceeds, enlarging as it goes to the sewer beneath the railroad track. The sanitary condition is not materially changed; the sewage-drains remain in the cellars; there is no sewer in the adjoining part of Thirteenth street; the soil water is not lowered; the slop-water, as before, covers the bricks; no longer soaking into our well, rather more of it must go into the cellars. So little of either good or harm has resulted, that the whole thing is a failure, except as the discussion may have done some good. (Fig. 42.)

(56) In August, 1884, it seemed well to call the attention of the Board to a sanitary nuisance infecting a block of houses. The drainage is by eight-inch pipe beneath the front parlors; each house has its drain connections with this main by **T** joints, all buried in the cellar floor. The plan is neat; the tenants like it. The proprietor of the grocery, 1520, reports that his cellar had water in it, and things floating about; workmen discovered that the drain passes into the next cellar, 1522, and dug it up with its dirt. There is little if any difference of opinion among builders and plumbers about the salubrity of the sewage-drain in the cellar; they all like it; the sanitary inspector cannot see it; thousands of new houses are now building with these leaky receptacles of sewage filth hid away in the cellars.

(57) We must get rid of the nuisance; we need not go over the

evidence that this arrangement is unwholesome. I own one of this block of houses, and have known the occupants for five years. The average population is six, the deaths were three ; this is a yearly death rate of 10 per cent., or 100 per M ; while the death rate of the city was 21 per M. The house is occupied by a family of four, the remnant of a larger family.

As for remedies, we have to deal with the owners of the houses. In the deed of each house is a clause reserving right of way for this drain. If the owners can be brought to reason, the drain can be removed from the cellars to the gardens without inconvenience. The expense would be inconsiderable.

This report was before the Board. The place was visited by the sanitary inspector. A sewer was built in front of the houses. The soil water is lower, the cellars drier ; there is probably less sickness. But the sewage drain remained in the cellars.

(58) All went nicely till the cyclone of August 3d, 1885. On that day four inches of rain fell, and backed into the cellars in a way to astonish the folks. One gentleman reported that the water in his cellar was a yard deep.

" You may be suspected of having seen the water with big eyes ; did you measure it ? "

" No ; but I noticed the marks it made on the steps. It was three steps and a half."

"Three and a half, at eight inches ; say, then, twenty-eight inches."

" It was deeper ; the steps must be nine inches. It was thirty-two inches, fully."

" Well, practically, it makes little difference, 28 or 32, 36 or 40. There was a good deal of water, and some work to get it out."

" Not much work. It ran out so fast that it was all gone by noon next day, except the filth. A girl with a bucket and a broom made all snug in two or three hours."

This water ran in from the sewer, and was fairly distributed in all the cellars. Eight blocks of houses, east, west, north and south, were similarly flooded. The water must have come from the street pavement, and entered the sewer at the inlet, on this corner. One of the residents reported the nuisance and annoyance.

(59) August 14th, 1885. An order of the Board was served on all the premises " to have removed *immediately* a certain nuisance . . . arising from a defective drain through cellar." The " defective drain " is a nuisance, but its only defect is that it is in the cellar.

We suggested to the owners to transfer the drain to the gardens; and two assented, so that we have three draining through the gardens (Figs. 42 and 43). Four were already at work connecting their old sewage-drains with the sewer in front. Two eventually followed suit in the same direction, in strict accordance with the order of the Board. The last one, being above the middle of the row, had to go to the sewer in front, but in his cellar he used iron soil-pipe attached to the cellar wall. So it stands (Fig. 43).*

(60) For this block of houses, what have we gained by stirring this dirt? What have we lost?

The agitation has given us a sewer the whole length of the block. This is a good thing for health, even if not needed for sewage. It is an artificial drain for the soil, ten feet deep, thus lowering the soil water, keeping the soil dry, the cellars dry. The agitation has enabled us to build a drain across three gardens. This lowers the soil water on each side and beyond for fifty feet, thus improving the sanitary condition of nine houses, six of our own block, and three of the adjoining block on Sixteenth street. Drains of this kind were made for this purpose before there was any thought of getting rid of dirt in this way; and they are still needed. The agitation has enabled us to dig up and throw away the sewage-drains from four cellars. The agitation has caused six owners to waste money in aggravating the nuisance of sewage-drains. By directing the plumbers to follow blindly the orders of the Board, they have increased the quantity of these leaky receptacles of filth hidden in the cellars. The increased number of openings to the sewer are a matter of indifference. With one opening for the whole block the water arose about as fast in the cellars as in the sewer. Perhaps the water will run out a little faster by the six openings, but as it ran out before in about twenty-four hours, this gain is not worth counting. The only valuable returns that these men have for their money is a lesson, which they seem greatly to have needed. They want to improve the sanitary condition of their property; they get useful information from such incidents as this. We have gained much, we have lost next to nothing; and besides, our neighbors seeing our good work, some of them have corrected their drainage on our plan; so that, considering the amount of hammering commonly needed to accomplish any reform, the result is, on the whole, a gratifying success.

* An employé of the Board, having inspected the work, affected astonishment that, all having been ordered to drain to the front, 1534 and 1536 were drained through the gardens. He suggested that the work would have to be done over again.

Fig. 43

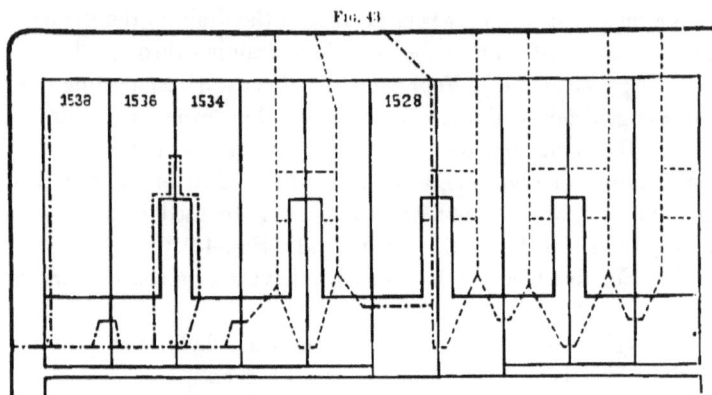

| 1538 | 1536 | 1534 | | | 1528 | | |

Drainage, as improved by the recent orders.

(61) We need an efficient Board of Health; such a Board may be composed of men or women, or both, but they cannot be ignorant of the subjects with which they have to deal. We would not needlessly criticise the present Board—the Board of Nuisances—the Board that hides leaky receptacles of filth in cellars. This Board is composed of Christian gentlemen; they devote their time and energies to the service of the people; and, as the rule, without other reward than the gratification of an impulse to love their neighbor: they "love even their enemies." We think well of the lawyer, but the cobbler is not so inefficient as he in the matter of mending shoes; we do not disparage the shoemaker, formerly Justice of Peace, when we suggest that the Magistrate, knowing more about law, does the business better. The plumber can make a wipe-joint, and his work in this way is perfect, notwithstanding some absurd and clamorous fault-finding. As a rule, "the cobbler should stick to his last;" but Bunyan, with all his tinkering, found time to read his Bible, and wrote "Pilgrim's Progress." Franklin did not muddle his brain with beer, and thus saved money to buy many books, found time to study them, and eventually he flew a kite. Faraday, an industrious blacksmith, found time to study, and took part with Ampere and Henry, in developing electro-magnetism into a science which, with the help of the mechanicians, is revolutionizing the world. These are, perhaps, exceptions; but no proverb could obstruct such workers as these, could do them no harm.

(62) Our ideal "Board of Health" is principally composed of persons who have made Hygiene a matter of study. The legislation necessary for such an organization must be discussed by our lawyers and law-makers; and there is good reason to hope for great improvement in the near future.

INDEX.

73

CATALOGUE
No. 1.

CATALOGUE

OF

MEDICAL, DENTAL,

Pharmaceutical & Scientific Publications,

WITH A CLASSIFIED INDEX,

PUBLISHED BY

P. BLAKISTON, SON & CO.,

(SUCCESSORS TO LINDSAY & BLAKISTON)

Booksellers, Publishers and Importers of Medical and Scientific Books,

No. 1012 WALNUT STREET, PHILADELPHIA.

THE FOLLOWING CATALOGUES WILL BE SENT FREE TO ANY ADDRESS, UPON APPLICATION.

This Catalogue, including all of our own publications.

A Catalogue of Books for Dental Students and Practitioners.

A Catalogue of Books on Chemistry, Pharmacy, The Microscope, Hygiene, Human Health, Sanitary Science, Technological Works, etc.

Students' Catalogue, including the "Quiz-Compends" and some of the most prominent **Text-books** and **manuals** for medical students.

A Complete Classified Catalogue (32 pages) of all Books on Medicine, Dentistry, Pharmacy and Collateral Branches. English and American.

A Catalogue of Medical and Scientific Periodicals and Physicians' Visiting Lists, giving club rates.

P. Blakiston, Son & Co.'s publications may be had through Booksellers in all the principal cities of the United States and Canada, or any book will be sent, postpaid, by the publishers, upon receipt of price, or will be forwarded by express, C. O. D., upon receiving a remittance of 25 per cent. of the amount ordered, to cover express charges. Money should be remitted by postal note, money order, registered letter, or bank draft.

☞ All new books received as soon as published. Special facilities for importing books from England, Germany and France.

CLASSIFIED LIST, WITH PRICES,

OF ALL BOOKS PUBLISHED BY

P. BLAKISTON, SON & CO., PHILADELPHIA.

☞ For further information in reference to these Books, send for full descriptive catalogue (No. 1), which will be sent, free, to any address. When the price is not given below, the book is not to be had at present. Cloth binding, unless otherwise specified.

ANÆSTHETICS.

Sansom. Chloroform. -	$1.25
Turnbull. 2d Ed. -	1.50
—— Cocaine. - -	.50

ANATOMY.

Handy. Text-book. -	3.00
Heath. Practical. -	5.00
Holden. Dissections. -	5.00
—— Landmarks. -	
Morris. On the Joints. -	5.50
Potter. Compend of -	1.00
—— Visceral. -	1.00
Wilson. 10th Ed. -	6.00

ATLASES AND DIAGRAMS.

Bentley and Trimens. Medicinal Plants. - - -	75.00
Braune. Of Anatomy. -	8.00
Flower. Of Nerves. -	3.50
Fox. Of Skin Dis. -	20.00
Godlee. Of Anatomy. -	20.00
Heath. Operative Surgery.	12.00
Hutchinson. Surgery. -	25.00
Jones. Membrana Tympani.	4.00
Marshall's Physiol. Plates.	80.00
Schultze. Obstetrical Plates.	25.00

BRAIN AND INSANITY.

Bucknill and Tuke. Psychological Medicine. - -	8.00
Gowers. Diagnosis of Diseases of the Brain. - - -	2.00
Mann's Psychological Med.	5.00
Roberts, Surgery of -	1.25
Wood. Brain and Overwork	.50

CHEMISTRY.

Allen. Commercial Analysis. Volume I. - - -	4.50
Bartley. Medical. -	2.50
Bloxam's Text-Book. -	3.75
—— Laboratory. -	1.75
Bowman's Practical. -	2.00
Frankland. How to Teach.	
Kollmyer. Key to. -	2.25
Leffmann's Compend. -	1.00
Muter. Med'l and Pharm.	6.00
—— Practical and Analy.	2.50
Richter's Inorganic. -	2.00
—— Organic. -	3.00
Stammer. Problems. -	.75
Sutton. Volumetric Anal.	5.00
Thompson's Physics. -	
Trimble. Analytical. -	1.50
Vacher's Primer of. -	.50
Valentin. Quant. Analy.	3.00
Ward's Compend of. -	1.00
Watts. Physical and Inorg.	2.25
Wolff. Applied Medical Chemistry. - - -	1.50

CHILDREN.

Chavasse. Mental Culture of.	1.00
Day. Diseases of. - -	3.00
Dillnberger. Women and.	1.50
Ellis. Manual of Dis. of.	3.00
—— Mother's book on.	.75
Goodhart and Starr. -	3.00
Hale. Care of. - -	.75
Hillier. Diseases of. -	1.25
Meigs. Infant Feeding and Milk Analysis, -	
Meigs and Pepper's Treatise.	6.00
Smith. Wasting Diseases of.	3.00
Starr, Digestive Organs of,	

COMPENDS

And The Quiz-Compends.

Brubaker's Physiology.	$1.00
Horwitz. Surgery. -	1.00
Hughes. Practice. 2 Pts. Ea.	1.00
Landis. Obstetrics. -	1.00
Leffmann's Chemistry. -	1.00
Mendenhall's Vade Mecum.	
Potter's Anatomy. -	1.00
—— Visceral Anatomy	1.00
—— Materia Medica. -	1.00
Roberts. Mat. Med. and Phar.	2.00
Stewart, Pharmacy. -	1.00
Ward's Chemistry. -	1.00

DEFORMITIES.

Churchill. Face and Foot.	3.50
Coles. Of Mouth. - -	4.50
Prince. Orthopædics. -	4.50
Reeves. " -	2.25

DENTISTRY.

Barrett. Dental Surg. -	1.00
Coles. Dental Note Book.	1.00
Flagg. Plastics. -	4.00
Gorgas. Dental Medicine.	3.00
Harris. Principles and Prac.	6.50
—— Dictionary of. -	6.50
Heath. Dis. of Jaws. -	4.50
Hunter. Mechanical Dent.	1.50
Leber and Rottenstein. Caries. - - - -	1.25
Richardson. Mech. Dent.	4.00
Stocken. Materia Medica.	2.50
Taft. Operative Dentistry.	4.25
——, Index of Dental Lit.,	
Tomes. Dental Surgery, -	
—— Dental Anatomy.	4.25
White. Mouth and Teeth.	.50

DICTIONARIES.

Cleveland's Pocket Medical.	.75
Cooper's Surgical. - -	12.00
Harris' Dental. - -	6.50
Longley's Pronouncing -	1.00

DIRECTORY.

Medical, of Philadelphia, Pa., Del. and South N. J.	2.50

EAR.

Burnett. Hearing, etc.	.50
Dalby. Diseases of. -	1.50
Jones. Aural Surgery. -	2.75
—— Membrana Tympani.	4.00
Woakes. Deafness, etc.	1.50
—— Catarrh, etc. -	

ELECTRICITY.

Althaus. Medical Electricity.	6.00
Reynolds. Clinical Uses.	1.00

EYE.

Arlt. Diseases of. -	2.50
Carter. Eyesight. -	1.25
Daguenet. Ophthalmoscopy.	1.50
Fenner. Vision. -	3.50
Gowers. Ophthalmoscopy.	6.00
Harlan. Eyesight. -	.50
Higgins. Handbook. -	.50
Jones. Sight and Hearing.	.50
Liebreich. Atlas of Ophth.	15.00
Macnamara. Diseases of.	4.00
Morton. Refraction. -	1.00
Wolfe. Diseases of. -	7.00

FEVERS.

Welch. Enteric Fever. -	2.00

HEADACHES.

Day. Their Treatment, etc.	$1.25
Wright. Causes and Cure.	.50

HEALTH AND DOMESTIC MEDICINE.

Bulkley. The Skin. -	.50
Burnett. Hearing. -	.50
Cohen. Throat and Voice.	.50
Dulles. Emergencies. -	.75
Harlan. Eyesight. -	.50
Hartshorne. Our Homes.	.50
Hufeland. Long Life. -	1.00
Lincoln. Hygiene. -	.50
Osgood. Winter. - -	.50
Packard. Sea Air, etc.	.50
Richardson's Long Life.	.50
Tanner. On Poisons. -	.75
White. Mouth and Teeth.	.50
Wilson. Summer. -	.50
Wilson's Domestic Hygiene.	1.00
Wood. Brain Work. -	.50

HEALTH RESORTS.

Madden. Foreign. -	2.50
Packard. Sea Air and Bath'g.	.50
Solly. Colorado Springs.	.25
Wilson. The Ocean as a.	2.50

HEART.

Balfour. Diseases of. -	5.00
Fothergill. Diseases of.	3.50
Sansom. Diseases of. -	1.25

HISTOLOGY.

See Microscope and Pathology.

HOSPITALS.

Burdett. Cottage Hospitals.	4.50
—— Pay Hospitals. -	2.25
Domville. Hospital Nursing.	.75

HYGIENE.

Bible Hygiene. -	1.00
Frankland. Water Analysis.	1.00
Fox. Water, Air, Food.	4.00
Lincoln. School Hygiene.	.50
Parke's Hygiene. 6th Ed.	3.00
Wilson's Handbook of. -	2.75
—— Domestic. - -	1.00
—— Drainage. - -	1.00
—— Naval. - -	3.00

KIDNEY DISEASES.

Edwards. How to Live with Bright's Disease. - -	.50
Greenhow. Addison's Dis.	3.00
Ralfe. Dis. of Kidney, etc.	2.75
Tyson. Bright's Disease.	3.50

LIVER.

Habershon. Diseases of.	1.50
Harley. Diseases of. -	3.00

LUNGS AND CHEST.

See Phy. Diagnosis and Throat.

MARRIAGE.

Ryan. Philosophy of. -	1.00
Walker. For Beauty, Health,	1.00

MATERIA MEDICA.

Biddle. 9th Ed. - -	4.00
Charteris. Manual of. -	
Gorgas. Dental. - -	3.00
Merrell's Digest. - -	4.00
Potter's Compend of. -	1.00
—— Handbook of. -	
Roberts' Compend of. -	2.00

MEDICAL JURISPRUDENCE.

Abercrombie's Handbook, $2.50
Reese's Text-book of - 4.00
Woodman and Tidy's Treatise, including Toxicology. 7.50

MICROSCOPE.

Beale. How to Work with. 7.50
—— In Medicine. - 7.50
Carpenter. The Microscope. 5.50
Lee. Vade Mecum of. - 3.00
MacDonald. Examination of Water by. - - - 2.75
Martin. Mounting. - 2 75
Wythe. The Microscopist. 3.00

MISCELLANEOUS.

Allen. The Soft Palate. .50
Beale. Life Theories, etc. 2.00
—— Slight Ailments. 1.25
Black. Micro-Organisms. 1.50
Cobbold. Parasites, etc. 5.00
Edwards. Malaria. - .50
—— Vaccination. - .50
—— Constipation. - ——
Gross. Life of Hunter. 1.25
Hardwicke. Med. Educat'n. 3.00
Hare. Tobacco. Paper, .50
Hodge. Fœticide. - .50
Holden. The Sphygmograph. 2.00
Kane. Opium Habit. - 1.25
MacMunn. The Spectroscope 3.00
Mathias. Legislative Manual. .50
Sieveking. Life Insurance. 2.00
Smith. Ringworm - 1.00
Smythe. Med'l Heresies. 1.25
Wickes. Sepulture. - 1.50

NERVOUS DISEASES.

Buzzard. Ner. Affections. 5.00
Flowers. Atlas of Nerves. 3.50
Gowers. Dis. of Spinal Cord. 1.50
—— Epilepsy. - - 4.00
Granville. Nerve Vibration. 2.00
Page. Injuries of Spine. 3.50
Radcliffe. Epilepsy, Pain, etc. 1.25
Tuke. Hypnotism, etc. 1.75
Wilkes. Nervous Diseases. 6.00

NURSING.

Cullingworth. Manual of. 1.00
—— Monthly Nursing. .50
Domville's Manual. - .75
Record for the Sick Room. .25
Temperature Charts. - .50

OBSTETRICS.

Barnes. Obstetric Operations. 3.75
Lazeaux and Tarnier. New Ed. Colored Plates. - ——
Gallabin's Manual of. - ——
Glisan's Text-book. - 4.00
Landis. Compend. - 1.00
Meadows. Manual. - 2.00
Rigby and Meadow's. - .50
Savage. Female Pelvic Org. 12.00
Schultze. Diagrams. - 25.00
Swayne's Aphorisms - 1.25

OSTEOLOGY.

Holden's Text-book. 1 Vol. 6.00

PATHOLOGY & HISTOLOGY.

Gibbes. Practical - - 1.50
Gilliam. Essentials of. - 2.00
Jones and Sieveking. - 5.50
Paget's Surgical Path. - 7.00
Piersol. Histology, 40 Plts. 6.50
Kindfleisch. General. 2.00
Virchow. Post-mortems. 1.00
Wilkes and Moxon. - 6.00

PHARMACY.

Beasley's Druggists' Rec'ts. 2.25
—— Formulary. - - 2.25
Flückiger. Cinchona Barks. 1.50
Kirby. Pharm. of Remedies. 2.25
Mackenzie. Phar. of Throat. 1.00
Merrell's Digest. - 4.00
Oldberg. Unofficial Pharm. 3.50
Piesse. Perfumery. - 5.50
Proctor. Practical Pharm. 4.50

PHYSICAL DIAGNOSIS.

Roberts. Compend of. $2.00
Stewart's Compend. - 1.00
Sweringen's Pharm. Lex. 3.00
Tuson. Veterinary Pharm. 2.50

Barth and Roger. - 1.00
Bruen's Handbook. 2d Ed. 1 50
West. Exam. of Chest. 1.75

PHYSIOLOGY.

Beale's Bioplasm. - - 2.25
—— Protoplasm. - - ——
Brubaker's Compend. - 1.00
Fulton's Text-book. - 4.00
Kirkes' 11th Edition.
  Cloth, 4.00; Sheep, 5.00
Landois. Text-book. 2 Vols. 10.00
Sanderson's Laboratory B'k. 5.00
Tyson's Cell Doctrine. - 2.00
Yeo's Student's Manual - 4.00

POISONS.

Black. Formation of. - 1 50
Reese. Toxicology. - 4.00
Tanner. Memoranda of. .75

PRACTICE.

Aitken. 2 Vols. New Ed. 12.00
Beale. Slight Ailments. 1.25
Charteris. Handbook of. 1.25
Cormack. Clinical Lectures. 5.00
Fagge's Practice. 2 Vols. ——
Fenwick's Outlines of. - 1.25
Hughes. Compend of. 2 Pts. 2.00
Roberts. Text-book. 5th Ed. 5.00
Tanner's Index of Diseases. 3.00
Warner's Case Taking. 1.75

PRESCRIPTION BOOKS.

Beasley's 3000 Prescriptions. 2.25
—— Receipt Book. - 2.25
—— Formulary. - 2.25
Oldberg's New Prescriptions. 1.25
Pereira's Pocket-book. - 1.00
Wythe's Dose and Symptom Book. - - - - 1.00

RECTUM AND ANUS.

Allingham. Diseases of. 1.25
Cripps. Diseases of. - 4.50

SKIN AND HAIR.

Bulkley. The Skin, - .50
Cobbold. Parasites. - 5.00
Fox. Atlas of Skin Dis. 20.00
Van Harlingen. Diagnosis and Treatment of Skin Dis. 1.75
Wilson. Skin and Hair. 1.00

STIMULANTS & NARCOTICS.

Anstie. On. - - 3.00
Kane. Opium Habit, etc. 1.25
Lizars. On Tobacco. - .50
Miller. On Alcohol - .50
Parrish. Inebriety. - 1.25

STOMACH & INDIGESTION.

Allbutt. Visceral Neuroses. 1.50
Edwards. Constipation ——
Fenwick. Atrophy of. - 3.25
Gill. Indigestion. - 1.25

SURGERY.

Cooper's Dictionary of. - 12.00
Druitt's Handbook of. - ——
Gamgee. Wounds and Fractures. - - - - 3.50
Heath's Operative. - 12.00
—— Minor, - - ——
—— Surgical Diagnosis. 1 25
—— Diseases of Jaws. 4.50
Horwitz. Compend. 2d Ed. 1.00
Hutchinson's Clinical - 25.00
Mears. Practical. 3 75
Pye. Surgical Handicraft. 5.00
Roberts. Surgical Delusions. .50
Watson's Amputations. 5.50

TECHNOLOGICAL BOOKS.

See also Chemistry.

Gardner. Brewing, etc. 1.75
—— Bleaching & Dyeing. 1.75

THERAPEUTICS.

Gardner. Acetic Acid, etc. $1.75
Overman. Mineralogy. 1 00
Piesse. Perfumery, etc. 5.50
Piggott. On Copper. - 1.00
Thompson. Physics. - ——

Cohen. Inhalations. - 1.25
Headland. Action of Med. 3.00
Kirby. Selected Remedies. 2.25
Mays. Therap. Forces. 1.25
Ott. Action of Medicines. 2.00
Potter's Compend. - 1.00
Waring's Practical. - ——

THROAT AND VOICE.

Cohen. Throat and Voice. .50
—— Inhalations. - 1.25
Dobell. Winter Cough, etc. 3.50
Greenhow. Bronchitis. 1.25
Holmes. Laryngoscope. 1.00
James. Sore Throat - 1.25
Mackenzie. Throat & Nose. 6.00
—— Larynx. - 1.25
—— Hay Fever. - .50
—— Pharmacopœia. - 1.25
Potter. Defects of Speech. 1.00
Thorowgood. Asthma. 1.50

TRANSACTIONS AND REPORTS.

Penna. Hospital Reports. 1.25
Power and Holmes' Reports. 1.25
Trans. College of Physicians. 3.00
—— Amer. Surg. Assoc. 4.00

TUMORS AND CANCER.

Hodge. Note-book for. .50
Thompson. Of the Bladder. 1.75
Wells. Ovarian and Uterine. 7.00
—— Abdominal - 1.50

URINE & URINARY ORGANS.

Acton. Repro. Organs. 2.00
Beale. Urin. & Renal Dis. 1.75
—— Urin. Deposits. Plates. 2.00
Curling. On the Testes. 5.50
Legg. On Urine. - .75
Marshall and Smith. Urine. 1.00
Thompson. Urinary Organs. 1.25
—— Surg. of Urin. Organs. 1.25
—— Calculous Dis. - 1.00
—— Lithotomy. - 3 50
—— Prostate. - Paper, .25
—— Tumors of Bladder. 1.75
—— Stricture. - 2.00
Tyson. Exam. of Urine. 1.50

VENEREAL DISEASES.

Cooper. Syphilis. - - 3.50
Durkee. Gonorrhœa. - 3.50
Hill and Cooper's Manual. - ——
Lewin. Syphilis. - - 1.25

VETERINARY PRACTICE.

Armatage's Pocket-book of. 1.25
Tuson's Vet. Pharmacopœia. 2.50

VISITING LISTS AND ACCOUNT BOOKS.

Lindsay and Blakiston's Regular Edition. 1.00 to 3.00
—— Perpetual Edition. 1.25
Watson's Led. and Cash Bk. 4.00

WATER.

Fox. Water, Air, Food. 4.00
Frankland. Analysis of. 1.00
MacDonald. " " 2.75

WOMEN, DISEASES OF.

Byford's Text-book. - 5.00
—— Uterus. - 1.25
Courty. Uterus, Ovaries, etc. 6.00
Dillnberger. and Children. 1 50
Duncan. Sterility. - 2.00
Gallabin. Diseases of. - ——
Savage. Surgery of Female Pelvic Organs. - 12.00
Tilt. Change of Life. - 1.25

New Books—Just Ready.

HUGHES. A COMPEND OF THE PRACTICE OF MEDICINE. Physicians' Edition, in One Volume. Enlarged by a Section on Diseases of the Skin, and a complete Index. By DAN'L E. HUGHES, M.D., Demonstrator of Clinical Medicine in the Jefferson Medical College, Philadelphia; Fellow of the College of Physicians of Philadelphia. 12mo. Full Morocco, $2.50

*** This edition has been prepared specially for physicians, from the second revised and enlarged students' edition, which it exceeds somewhat in size. The very favorable reception which has been accorded it in its original form, leads the author and publishers to believe that in this shape its increased usefulness to physicians will be recognized. It contains the Synonyms, Definitions, Causes, Symptoms, Prognosis, Treatment, etc., of each disease. including many new prescriptions hitherto unpublished. It has been compiled from lectures of prominent professors, and in its preparation the author has consulted the latest writings of Professors Flint, Roberts, DaCosta, Reynolds, Loomis, Bartholow and others. It therefore contains the latest teachings and methods of treatment, arranged in the most concise, practical way, and contains information nowhere else collected in such a useful way.

MEIGS. MILK ANALYSIS AND INFANT FEEDING. A Practical Treatise on the Examination of Human and Cows' Milk, Cream, Condensed Milk, etc., and Directions as to the Diet of Young Infants. By ARTHUR V. MEIGS, M.D., Physician to the Pennsylvania Hospital and to the Children's Hospital; Fellow of the College of Physicians of Philadelphia, etc. 12mo.
Cloth, $1.00

VIRCHOW. POST-MORTEM EXAMINATIONS, with Especial Reference to Medico-Legal Practice. By PROF. RUDOLPH VIRCHOW, of the Berlin Charité Hospital. Translated by T. P. SMITH, M.D., Member of the Royal College of Surgeons, of England. With additional notes and new plates, from the Fourth German Edition. 4 Lithographic Plates. 12mo. Cloth, $1.00

MEARS' PRACTICAL SURGERY. Second Edition. Including Surgical Dressings, Bandaging, Fractures, Dislocations, Ligature of Arteries, Amputations, and Excisions of Bones and Joints. By J. EWING MEARS, M.D., Lecturer on Practical Surgery and Demonstrator of Surgery in Jefferson Medical College, Philadelphia. Second Edition. Revised and Enlarged. With 490 Illustrations. 12mo. 794 pages. Cloth, $3.75; Leather, $4.75

HABERSHON. DISEASES OF THE LIVER. On the Pathology and Treatment of some Diseases of the Liver. By S. O. HABERSHON, M.D., Lond., F.R.C.P., late Senior Physician to, and Lecturer on Medicine at, Guy's Hospital, London. Second Edition, Revised. 12mo. Cloth, $1.50

HARE. ON TOBACCO. The Physiological and Pathological Effects of the Use of Tobacco. By HOBART AMORY HARE, M.D., B.SC. Being the Fiske Fund Prize Dissertation for 1885. Illustrated. Octavo. Paper covers, 50 cts.

P. BLAKISTON, SON & CO., 1012 Walnut St., Philadelphia.

P. BLAKISTON, SON & CO.'S

Medical and Scientific Publications,

No. 1012 Walnut St., Philadelphia.

ABERCROMBIE. **Medical Jurisprudence,** for Medical and Legal Students and Practitioners. By John Abercrombie, M.D. 387 pages. Cloth, $2.50

ACTON. **The Functions and Disorders of the Reproductive Organs** in Childhood, Youth, Adult Age and Advanced Life, considered in their Physiological, Social and Moral Relations. By William Acton, M.D., M.R.C.S. Sixth Edition. 8vo. Cloth, $2.00

AITKEN. **Science and Practice of Medicine.** By William Aitken, M.D., F.R.S., Professor of Pathology in the Army Medical School, London. Seventh Edition. Revised throughout. 196 Engravings on Wood, and a Map. 2 vols. 8vo. Cloth, $12.00 ; Leather, $14.00

ALLBUTT. **Visceral Neuroses.** On Neuralgia of the Stomach, and Allied Disorders. By T. Clifford Allbutt, M.D., F.R.S., Consulting Physician to the Leeds General Infirmary. 8vo. Cloth, $1.50

ALLEN. **Commercial Organic Analysis.** A Treatise on the Modes of Assaying the Various Organic Chemicals and Products employed in the Arts, Manufactures, Medicine, etc., with Concise Methods for the Detection of Impurities, Adulterations, etc. Second Edition. Revised and Enlarged. By Alfred Allen, F.C.S.
Vol. I. Alcohols, Ethers, Vegetable Acids, Starch and its Isomers, etc.
$4.50
Vol. II. Fixed Oils and Fats, Hydrocarbons and Mineral Oils, Phenols and their Derivatives, Coloring Matters, etc. *In Press.*
Vol. III. Cyanogen Compounds, Alkaloids, Animal Products, etc. *In Press.*

ALLEN'S New Method of Recording the Motions of the Soft Palate. By Harrison Allen, M.D., Professor of Physiology University of Pennsylvania. Cloth, .50

ALLINGHAM. **Diseases of the Rectum.** Fistula, Hæmorrhoids, Painful Ulcer, Stricture, Prolapsus, and other Diseases of the Rectum, their Diagnosis and Treatment. By William Allingham, F.R.C.S. Fourth Edition, Enlarged.
Illustrated. 8vo. Paper covers, .75 ; Cloth, $1.25
London Edition, thick paper and larger type, $2.00

ALTHAUS. **Medical Electricity.** Theoretical and Practical. Its Use in the Treatment of Paralysis, Neuralgia, and other Diseases. By Julius Althaus, M.D. Third Edition, Enlarged. 246 Illustrations. 8vo. Cloth, $6.00

ANSTIE. **Stimulants and Narcotics.** With special researches on the Action of Alcohol, Ether and Chloroform on the Vital Organism. By Francis E. Anstie, M.D. 8vo. Cloth, $3.00

ARLT. **Diseases of the Eye.** Clinical Studies on Diseases of the Eye. Including the Conjunctiva, Cornea and Sclerotic, Iris and Ciliary Body. By Dr. Ferd. Ritter von Arlt, University of Vienna. Authorized Translation by Lyman Ware, M.D., Surgeon to the Illinois Charitable Eye and Ear Infirmary, Chicago.
Illustrated. 8vo. Cloth, $2.50

ARMATAGE. **The Veterinarian's Pocket Remembrancer.** Containing concise directions for the Treatment of Urgent or Rare Cases, embracing Semeiology, Diagnosis, Prognosis, Surgery, Therapeutics, etc. 32mo. Cloth, $1.25

BALFOUR. **Clinical Lectures on Diseases of the Heart and Aorta.** By G. W. Balfour, M.D. Illustrated. Second Edition. Cloth, $5.00

BARNES. **Lectures on Obstetric Operations,** including the Treatment of Hemorrhage, and forming a Guide to Difficult Labor. By Robert Barnes, M.D., F.R.C.P. Fourth Edition. Illustrated. 8vo. Cloth, $3.75

5

BARRETT. Dental Surgery for General Practitioners and Students of Medicine and Dentistry. Extraction of Teeth, etc. By A. W. BARRETT, M.D. Illustrated.
Cloth, $1.00

BARTH AND ROGER. Auscultation and Percussion. 12mo. Cloth, $1.00

BARTLEY. Medical Chemistry. A Text-book for Medical and Pharmaceutical Students. By E. H. BARTLEY, M.D., Associate Professor of Chemistry at the Long Island College Hospital; President of the American Society of Public Analysts; Chief Chemist, Board of Health, of Brooklyn, N. Y. With Illustrations, Glossary and Complete Index. 12mo. Cloth, $2.50

BEALE. On Slight Ailments; their Nature and Treatment. By LIONEL S. BEALE, M.D., F.R.S., Professor of Practice, King's Medical College, London. Second Edition. Enlarged and Illustrated. Paper covers, .75; Cloth, $1.25
Finer Edition, Heavy Paper. Extra Cloth, $1.75

Urinary and Renal Diseases and Calculous Disorders. Hints on Diagnosis and Treatment. Demi-8vo. 356 pages. Cloth, $1.75

The Use of the Microscope in Practical Medicine. For Students and Practitioners, with full directions for examining the various secretions, etc., in the Microscope. Fourth Edition. 500 Illustrations. 8vo. Cloth, $7.50

How to Work with the Microscope. A Complete Manual of Microscopical Manipulation, containing a full description of many new processes of investigation, with directions for examining objects under the highest powers, and for taking photographs of microscopic objects. Fifth Edition, Containing over 400 Illustrations, many of them colored. 8vo. Cloth, $7.50

Protoplasm ; or Matter and Life. Sixteen Colored Plates. Part I. Dissentient. Part II. Demonstrative. Part III. Suggestive. *New Ed. Preparing.*

Bioplasm. A Contribution to the Physiology of Life, or an Introduction to the Study of Physiology and Medicine, for Students. With numerous Illustrations. Cloth, $2.25

Life Theories ; Their Influence upon Religious Thought. Six Colored Plates.
Cloth, $2.00

On Life and Vital Action in Health and Disease. 12mo. Cloth, $2.00

One Hundred Urinary Deposits, on eight sheets, for the Hospital, Laboratory, or Surgery. New Edition. 4to. Paper, $2.00

BEASLEY'S Book of Prescriptions. Containing over 3100 Prescriptions, collected from the Practice of the most Eminent Physicians and Surgeons—English, French and American ; a Compendious History of the Materia Medica, Lists of the Doses of all Officinal and Established Preparations, and an Index of Diseases and their Remedies. By HENRY BEASLEY. Sixth Edition. Revised and Enlarged. Cloth, $2.25

Druggists' General Receipt Book. Comprising a copious Veterinary Formulary; Recipes in Patent and Proprietary Medicines, Druggists' Nostrums, etc.; Perfumery and Cosmetics ; Beverages, Dietetic Articles and Condiments; Trade Chemicals, Scientific Processes, and an Appendix of Useful Tables. Ninth Edition. Revised. Cloth, $2.25

Pocket Formulary and Synopsis of the British and Foreign Pharmacopœias. Comprising Standard and Approved Formulæ for the Preparations and Compounds Employed in Medical Practice. Eleventh Edition. Cloth, $2.25

BENTLEY AND TRIMEN'S Medicinal Plants. A New Illustrated Work, containing full botanical descriptions, with an account of the properties and uses of the principal plants employed in medicine, especial attention being paid to those which are officinal in the British and United States Pharmacopœias. The plants which supply food and substances required by the sick and convalescent are also included. By R. BENTLEY, F.R.S., Professor of Botany, King's College, London, and H. TRIMEN, M.B., F.H.S., Department of Botany, British Museum. Each species illustrated by a colored plate drawn from nature. In forty-two parts. Eight colored plates in each part.
Price reduced to $1.50 per part, or the complete work handsomely bound in 4 volumes. Half Morocco, Gilt, $75.00.

BIBLE HYGIENE; or Health Hints. By a physician. Written to impart in a popular and condensed form the elements of Hygiene; showing how varied and important are the Health Hints contained in the Bible, and to prove that the secondary tendency of modern Philosophy runs in a parallel direction with the primary light of the Bible. 12mo. Cloth, $1.00

BIDDLE'S Materia Medica and Therapeutics. Tenth Edition. For the Use of Students and Physicians. By Prof. JOHN B. BIDDLE, M.D., Professor of Materia Medica in Jefferson Medical College, Philadelphia. The Tenth Edition, thoroughly revised, and in many parts rewritten, by his son, CLEMENT BIDDLE, M.D., Assistant Surgeon, U. S. Navy, and HENRY MORRIS, M.D., Demonstrator of Obstetrics in Jefferson Medical College, Fellow of the College of Physicians, of Philadelphia, etc. The Botanical portions have been curtailed or left out, and the other sections, on the Physiological action of Drugs, greatly enlarged.
Cloth, $4.00; Leather, $4.75

BLACK. Micro-Organisms. The Formation of Poisons by Micro-Organisms. A Biological study of the Germ Theory of Disease. By G. V. BLACK, M.D., D.D.S.
Cloth, $1.50

BLOXAM. Chemistry, Inorganic and Organic. With Experiments. By CHARLES L. BLOXAM, Professor of Chemistry in King's College, London, and in the Department for Artillery Studies, Woolwich. Fifth Edition. With nearly 300 Engravings. 8vo. Cloth, $3.75; Leather, $4.75

Laboratory Teaching. Progressive Exercises in Practical Chemistry. Intended for use in the Chemical Laboratory, by those who are commencing the study of Practical Chemistry. 4th Edition. 89 Illus. Cloth, $1.75

BOWMAN. Practical Chemistry, including analysis, with about 100 Illustrations. By Prof. JOHN E. BOWMAN. Eighth English Edition. Revised by Prof. BLOXAM, Professor of Chemistry, King's College, London. Cloth, $2.00

BRAUNE. Atlas of Topographical Anatomy. Thirty-four Full-page Plates, Photographed on Stone, from Plane Sections of Frozen Bodies, with many other illustrations. By WILHELM BRAUNE, Professor of Anatomy at Leipzig. Translated and Edited by EDWARD BELLAMY, F.R.C.S., Lecturer on Anatomy, Charing Cross Hospital, London. 4to. Cloth, $8.00; Half Morocco, $10.00

BRUBAKER. Physiology. A Compend of Physiology, specially adapted for the use of Students and Physicians. By A. P. BRUBAKER, M.D., Demonstrator of Physiology at Jefferson Medical College, Prof. of Physiology, Penn'a College of Dental Surgery, Philadelphia. Third Edition. Revised, Enlarged and Illustrated. "No. 4, ? Quiz-Compend Series?" 12mo. Cloth, $1.00
Interleaved for the addition of notes, $1.25

BRUEN. Physical Diagnosis. For Physicians and Students. By EDWARD T. BRUEN, M.D., Asst. Professor of Physical Diagnosis in the University of Pennsylvania. Illustrated by Original Wood Engravings. 12mo. 2d Ed. Cloth, $1.50

BUCKNILL AND TUKE'S Manual of Psychological Medicine: containing the Lunacy Laws, the Nosology, Ætiology, Statistics, Description, Diagnosis, Pathology (including morbid Histology) and Treatment of Insanity. By JOHN CHARLES BUCKNILL, M.D., F.R.S., and DANIEL HACK TUKE, M.D., F.R.C.P. Fourth Edition, much enlarged, with twelve lithographic and numerous other illustrations. 8vo. Cloth, $8.00

BULKLEY. The Skin in Health and Disease. By L. DUNCAN BULKLEY, M.D., Attending Physician at the New York Hospital. Illustrated. Cloth, .50

BURDETT'S Pay Hospitals and Paying Wards throughout the World. Facts in support of a rearrangement of the system of Medical Relief. By HENRY C. BURDETT, M.D. 8vo. Cloth, $2.25

Cottage Hospitals. General, Fever and Convalescent; their Progress, Management and Work. Second Edition. Rewritten and Enlarged, with Plans and Illustrations. Crown 8vo. Cloth, $4.50

BURNETT. Hearing, and How to Keep It. By CHAS. H. BURNETT, M.D., Prof. of Diseases of the Ear, at the Philadelphia Polyclinic. Illustrated. Cloth, .50

BUZZARD. Clinical Lectures on Diseases of the Nervous System. By THOS. BUZZARD, M.D. Illustrated. Octavo.   Cloth, $5.00

BYFORD. Diseases of Women. The Practice of Medicine and Surgery, as applied to the Diseases of Women. By W. H. BYFORD, A.M., M.D., Professor of Obstetrics and the Diseases of Women and Children, in the Chicago Medical College. Third Edition. Revised and Enlarged, much of it rewritten, with numerous additional illustrations. 8vo.   Cloth, $5.00; Leather, $6.00

    On the Uterus. The Chronic Inflammation and Displacement of the Unimpregnated Uterus. With Illustrations.   Paper, .75; Cloth, $1.25

CARPENTER. The Microscope and Its Revelations. By W. B. CARPENTER, M.D., F.R.S. Sixth Edition. Revised and Enlarged, with over 500 Illustrations and Lithographs.   Cloth, $5.50

CARTER. Eyesight, Good and Bad. A Treatise on the Exercise and Preservation of Vision. By ROBERT BRUDENELL CARTER, F.R.C.S. Second Edition, with 50 Illustrations, Test Types, etc. 12mo.   Paper, .75; Cloth, $1.25

CAZEAUX and TARNIER'S Midwifery. Seventh Revised and Enlarged Edition. With Colored Plates and numerous other Illustrations. The Theory and Practice of Obstetrics; including the Diseases of Pregnancy and Parturition, Obstetrical Operations, etc. By P. CAZEAUX, Member of the Imperial Academy of Medicine, Adjunct Professor in the Faculty of Medicine in Paris. Remodeled and rearranged, with revisions and additions, by S. TARNIER, M.D., Professor of Obstetrics and Diseases of Women and Children in the Faculty of Medicine of Paris. A New American, from the Eighth French and First Italian Edition. Edited and Enlarged by ROBERT J. HESS, M.D., Physician to the Northern Dispensary, Phila., etc. About 1100 pages quarto, with 12 Full-page Plates (five of which are beautifully colored) and over 175 Wood Engravings. Royal Square Octavo. *Sold by subscription only. Circulars and information will be sent, upon application to the Publishers.*

CHARTERIS. The Practice of Medicine. A Handbook. By M. CHARTERIS, M.D., Member of Hospital Staff and Professor in University of Glasgow. With Microscopic and other Illustrations.   Cloth, $1.25

    Materia Medica and Therapeutics. A Manual for Students.   *In Press.*

CHAVASSE. The Mental Culture and Training of Children. By PYE HENRY CHAVASSE. 12mo.   ,   Cloth, $1.00

CHURCHILL. Face and Foot Deformities. By FRED. CHURCHILL, M.D., Ass't Surgeon to the Victoria Hospital for Sick Children, London. Six Plain and Two Colored Lithographs. 8vo.   Cloth, $3.50

CLEAVELAND'S Pocket Dictionary. A Pronouncing Medical Lexicon, containing correct Pronunciation and Definition of terms used in medicine and the collateral sciences, abbreviations used in prescriptions, list of poisons, their antidotes, etc. By C. H. CLEAVELAND, M.D. Thirty-first Edition. 16mo.
    Cloth, .75; Tucks with Pocket, $1.00

COBBOLD. A Treatise on the Entozoa of Man and Animals, including some account of the Ectozoa. By T. SPENCER COBBOLD, M.D., F.R.S. With 85 Illustrations. 8vo.   Cloth, $5.00

COHEN on Inhalation, its Therapeutics and Practice, including a Description of the Apparatus Employed, etc. By J. SOLIS COHEN, M.D. With cases and Illustrations. A New Enlarged Edition. 12mo.   Paper, .75; Cloth, $1.25

    The Throat and Voice. Illustrated. 12mo.   Cloth, .50

COLES. Deformities of the Mouth, Congenital and Acquired, with Their Mechanical Treatment. By OAKLEY COLES, M.D., D.D.S. Third Edition. 83 Wood Engravings and 96 Drawings on Stone. 8vo.   Cloth, $4.50

    The Dental Student's Note-Book. A new Edition. 16mo.   Cloth, $1.00

COOPER'S Surgical Dictionary. A Dictionary of Practical Surgery and Encyclopædia of Surgical Science. By SAMUEL COOPER. New Edition. By SAMUEL A. LANE, F.R.C.S., assisted by various eminent Surgeons. 2 vols.   Cloth, $12.00

COOPER on Syphilis and Pseudo-Syphilis. By ALFRED COOPER, F.R.C.S., Surgeon to the Lock Hospital, to St. Marks, and to the West London Hospitals. Octavo. Cloth, $3.50

CORMACK'S Clinical Studies. Illustrated by Cases observed in Hospital and Private Practice. By Sir JOHN ROSE CORMACK, M. D., K. B., etc. Illustrated. 2 vols. 1127 pp. Cloth, $5.00

COURTY. The Uterus, Ovaries, etc. A Practical Treatise on Diseases of the Uterus, Ovaries and Fallopian Tubes. By Prof. A. COURTY, of Montpellier, France. Translated from the Third Edition, by his pupil and assistant, AGNES McLAREN, M.D., M.K.Q.C.P.I. With a Preface by J. MATHEWS DUNCAN, M.D., LL.D., F.R.S., Obstetric Physician to Saint Bartholomew's Hospital, London, With 431 Illustrations. 8vo. *Sold by Subscription.* Cloth, $6.00; Leather, $7.00

CRIPPS. Diseases of the Rectum and Anus, including a portion of the Jacksonian Prize Essay on Cancer. By HARRISON CRIPPS, M.D., Ass't Surgeon to St. Bartholomew's Hospital, London. Lithographic Plates and other Illustrations. Cloth, $4.50

CULLINGWORTH. A Manual of Nursing, Medical and Surgical. By CHARLES J. CULLINGWORTH, M.D., Physician to St. Mary's Hospital, Manchester, England. Second Edition. With 18 Illustrations. 12mo. Cloth, $1.00

A Manual for Monthly Nurses. 32mo. Cloth, .50

CURLING. On the Diseases of the Testis, Spermatic Cord and Scrotum. By T. B. CURLING, M.D., F.R.S. Fourth Edition, Enlarged and Illustrated. 8vo. Cloth, $5.50

DAGUENET'S Ophthalmoscopy. A Manual for the Use of Students. By Dr. DAGUENET. Translated from the French, by Dr. C. S. JEAFFERSON, F.R.C.S.E. Illustrated. 12mo. Cloth, $1.50

DALBY. The Ear. The Diseases and Injuries of the Ear. By W. B. DALBY, M.D., Surgeon and Lecturer on Aural Surgery, St. George's Hospital. With Illustrations. 12mo. Cloth, $1.50

DAY. Diseases of Children. A Practical and Systematic Treatise for Practitioners and Students. By Wm. H. DAY, M.D. Second Edition. Rewritten and very much Enlarged. 8vo. 752 pp. Price reduced. Cloth, $3.00; Sheep, $4.00

On Headaches. The Nature, Causes and Treatment of Headaches. Fourth Edition. Illustrated. 8vo. Paper, .75; Cloth, $1.25

DILLNBERGER. On Women and Children. A Handbook of the Treatment of the Diseases Peculiar to Women and Children. By Dr. EMIL DILLNBERGER. 12mo. Cloth, $1.50

DOBELL. On Winter Cough, Catarrh, Bronchitis, Emphysema, Asthma, etc. By HORACE DOBELL, M.D., Senior Physician to the Royal Hospital for Diseases of the Chest. Third Edition. Octavo. Cloth, $3.50

DOMVILLE. Manual for Nurses and others engaged in attending to the sick. By Ed. J. Domville, M.D. Fifth Ed. With Recipes for Sick-room Cookery. etc. Cloth, .75

DRUITT'S Modern Surgery. The Surgeon's Vade Mecum ; a Manual of Modern Surgery. By ROBERT DRUITT, F.R.C.S. Twelfth Enlarged Edition, with 369 Illustrations. 864 pp. *In Preparation.*

DULLES. What to Do First, In Accidents and Poisoning. By C. W. DULLES, M.D. Second Edition, Enlarged, with new Illustrations. Cloth, .75

DUNCAN, On Sterility in Women. By J. MATHEWS DUNCAN, M.D., LL.D., Obstetric Physician to St. Bartholomew's Hospital, etc. Octavo. Cloth, $2.00

DURKEE, On Gonorrhœa and Syphilis. By SILAS DURKEE, M.D. Sixth Edition. Revised and Enlarged, with Portrait and Eight Colored Illustrations. Cloth, $3.50

ELLIS. Diseases of Children. A Practical Manual of the Diseases of Children, with a Formulary. By EDWARD ELLIS, M.D. Late Physician to the Victoria Hospital for Children, London. Fourth Edition, Enlarged. Cloth, $3.00

What Every Mother Should Know. 12mo. Cloth, .75

*

EDWARDS. Bright's Disease. How a Person Affected with Bright's Disease Ought to Live. By Jos. F. EDWARDS, M.D. 2d Ed. Reduced to   Cloth, .50

Constipation. New Edition. Plainly Treated and Relieved Without the Use of Drugs. Second Edition. Price Reduced.   Cloth, .50

Malaria: What It Means; How to Escape It; Its Symptoms; When and Where to Look For It. Price Reduced.   Cloth, .50

Vaccination and Smallpox. Showing the Reasons in favor of Vaccination, and the Fallacy of the Arguments advanced against it, with Hints on the Management and Care of Smallpox patients.   Cloth, .50

The above four books bound in one volume.   Cloth, $1.50

FAGGE. The Principles and Practice of Medicine. By C. HILTON FAGGE, M.D., F.R.C.P., F.R.M.C.S., Examiner in Medicine, University of London; Physician to, and Lecturer on Pathology in, Guy's Hospital; Senior Physician to Evelina Hospital for Sick Children, etc. Arranged for the press by PHILIP H. PYE SMITH, M.D., Lect. on Medicine in Guy's Hospital. Including a section on Cutaneous Affections, by the Editor; Chapter on Cardiac Diseases, by SAMUEL WILKES, M.D., F.R.S., and Complete Indexes by ROBERT EDMUND CARRINGTON. Royal 8vo. 2 vols.
Sold only by subscription. Full information upon application to the Publishers.

FENNER. On Vision. Its Optical Defects, the Adaptation of Spectacles, Defects of Accommodation, etc. By C. S. FENNER, M.D. With Test Types and 74 Illustrations. Second Edition, Revised and Enlarged. 8vo.   Cloth, $3.50

FENWICK'S Outlines of Practice of Medicine. With Formulæ and Illustrations. By SAMUEL FENWICK, M.D., 12mo.   Cloth, $1.25

Atrophy of the Stomach and Its Effect on the Nervous Affections of the Digestive Organs. 8vo.   Cloth, $3.25

FLAGG'S Plastics and Plastic Filling; As Pertaining to the Filling of all Cavities of Decay in Teeth below Medium in Structure, and to Difficult and Inaccessible Cavities in Teeth of all Grades of Structure. By J. FOSTER FLAGG, D.D.S., Professor in the Philadelphia Dental College. 8vo. Second Ed.   Cloth, $4.00

FLOWER'S Diagrams of the Nerves of the Human Body. Exhibiting their Origin, Divisions and Connections, with their Distribution to the various Regions of the Cutaneous Surface, and to all the Muscles. By WILLIAM H. FLOWER, F.R.C.S., F.R.S., Hunterian Professor of Comparative Anatomy, and Conservator of the Museum of the Royal College of Surgeons. Third Edition, thoroughly revised. With six Large Folio Maps or Diagrams. 4to.   Cloth, $3.50

FLÜCKIGER. The Cinchona Barks Pharmacognostically Considered. By Professor FRIEDRICH FLÜCKIGER, of Strasburg. Translated by FREDERICK B. POWER, PH.D., Professor of Materia Medica and Pharmacy, University of Wisconsin. With 8 Lithographic Plates. Royal octavo.   Cloth, $1.50

FOTHERGILL. On the Heart and Its Diseases. With Their Treatment. Including the Gouty Heart. By J. MILNER FOTHERGILL, M.D., Member of the Royal College of Physicians of London. 2d Ed. Re-written. 8vo. Cloth, $3.50

FOX. Water, Air and Food. Sanitary Examinations of Water, Air and Food. By CORNELIUS B. FOX, M.D. 94 Engravings. 8vo.   Cloth, $4.00

FOX'S Atlas of Skin Diseases. Complete in Eighteen Parts, each containing Four Chromo-lithographic Plates, with Descriptive Text and Notes upon Treatment. In all 72 large colored plates. By TILBURY FOX, M.D., F.R.C.P., Physician to the Department for Skin Diseases in University College Hospital. 4to.
Each part $1.00; or 1 vol., Cloth, $20.00

FRANKLAND'S Water Analysis. For Sanitary Purposes, with Hints for the Interpretation of Results. By E. FRANKLAND, M.D., F.R.S. Illustrated. 12mo.
Cloth, $1.00

How to Teach Chemistry. Six Lessons to Science Teachers. Edited by G. G. CHALONER, F.C.S. Illustrated. Second Edition.   *In Press.*

FULTON'S Physiology. A Text-book. By J. FULTON, M.D., Professor at Trinity Medical College, Toronto. Second Edition, Illustrated. 8vo.   Cloth, $4.00

GALLABIN'S Midwifery. A Manual for Students and Practitioners. By A. LEWIS GALLABIN, M.D., F.R.C.P., Obstetric Physician to Guy's Hospital, London, and Professor of Midwifery in the same institution. Illustrated. *In Press*

GAMGEE. Wounds and Fractures. The Treatment of Wounds and Fractures. Clinical Lectures, by SAMPSON GAMGEE, F.R.S.E., Consulting Surgeon to Queen's Hospital, Birmingham. 34 Engravings. Second Edition. 8vo. Cloth, $3.50

GARDNER'S TECHNOLOGICAL SERIES. The Brewer, Distiller and Wine Manufacturer. A Handbook for all Interested in the Manufacture and Trade of Alcohol and Its Compounds. Edited by JOHN GARDNER, F.C.S. Illustrated.
Cloth, $1.75

    Bleaching, Dyeing, and Calico Printing. With Formulæ. Illustrated. $1.75

    Acetic Acid, Vinegar, Ammonia and Alum. Illustrated. Cloth, $1.75

GIBBES'S Practical Histology and Pathology. By HENEAGE GIBBES, M.B. 12mo. Third Edition. Cloth, $1.75

GILL. Indigestion: What it is; What it Leads to; and a New Method of Treating it. By JOHN BEADNELL GILL, M.D. Third Edition. 12mo. Cloth, $1.25

GILLIAM'S Pathology. The Essentials of Pathology; a Handbook for Students. By D. TOD GILLIAM, M.D., Professor of Physiology, Starling Medical College, Columbus, O. With 47 Illustrations. 12mo. Cloth, $2.00

GLISAN'S Modern Midwifery. A Text-book. By RODNEY GLISAN, M.D., Emeritus Professor of Midwifery and Diseases of Women and Children, in Willamette University, Portland, Oregon. 129 Illus. 8vo. Cloth, $4.00; Leather, $5.00

GODLEE'S Atlas of Anatomy. Illustrating most of the Ordinary Dissections and many not usually practiced by the Student. With References and an Explanatory Text, and 48 Colored Plates. By RICKMAN JOHN GODLEE, M.D., F.R.C.S. A large Folio Volume, with References, and a Separate Volume of Letter-press. The two Volumes, Atlas and Letter-press, Cloth, $20.00

GOODHART and STARR'S Diseases of Children. The Student's Guide to the Diseases of Children. By J. F. GOODHART, M.D., F.R.C.P., Physician to Evelina Hospital for Children, Demonstrator of Morbid Anatomy at Guy's Hospital. Edited, with notes and additions, by LOUIS STARR, M.D., Clinical Professor of Diseases of Children, in the University of Pennsylvania. Cloth, $3.00; Leather, $4.00

GORGAS'S Dental Medicine. A Manual of Materia Medica and Therapeutics. By FERDINAND J. S. GORGAS, M.D., D.D.S., Professor of the Principles of Dental Science, Dental Surgery and Dental Mechanism, in the Dental Department of the University of Maryland. Second Edition. Enlarged. 8vo. Cloth, $3.25

GOWERS, Spinal Cord. Diagnosis of Diseases of the Spinal Cord. With Colored Plates and Engravings. Third Edition. Enlarged. By WILLIAM R. GOWERS, M.D., Ass't Prof. Clinical Medicine, University College, London. Cloth, $1.50

    Ophthalmoscopy. A Manual and Atlas of Ophthalmoscopy. With 16 Colored Autotype and Lithographic Plates and 26 Wood Cuts, comprising 112 Original Illustrations of the Changes in the Eye in Diseases of the Brain, Kidneys, etc. 8vo. Cloth, $6.00

    Epilepsy and other Chronic Convulsive Diseases: Their Causes, Symptoms, and Treatment. 8vo. Cloth, $4.00

    Diagnosis of Diseases of the Brain. 8vo. Illustrated. Cloth, $2.00

GRANVILLE. Nerve Vibration and Excitation as Agents in the Treatment of Functional Disorder and Organic Disease. By J. MORTIMER GRANVILLE, M.D. Illustrated. 8vo. Cloth, $2.00

GROSS'S Biography of John Hunter. John Hunter and His Pupils. By S. D. GROSS, M.D., Professor of Surgery in Jefferson Medical College, Philadelphia. With a Portrait. 8vo. Paper, .75; Cloth, $1.25

GREENHOW. Chronic Bronchitis, especially as connected with Gout, Emphysema, and Diseases of the Heart. By E. HEADLAM GREENHOW, M.D. 12mo.
Paper, .75; Cloth, $1.25

GREENHOW. Addison's Disease. Illustrated by Plates and Reports of Cases. By E. HEADLAM GREENHOW, M.D. 8vo. Cloth, $3.00

HABERSHON. On Some Diseases of the Liver. By S. O. HABERSHON, M.D., F.R.C.P., late Senior Physician to Guy's Hospital. A New Edition. Cloth, $1.50

HALE. On the Management of Children in Health and Disease. A Book for Mothers. By AMIE M. HALE, M.D. New Enlarged Edition. 12mo. Cloth, .75

HANDY'S Text-Book of Anatomy and Guide to Dissections. For the Use of Students. By W. R. HANDY, M.D. 312 Illustrations. 8vo. Cloth, $3.00

HARDWICKE. Medical Education and Practice in All Parts of the World. Containing Regulations for Graduation at the Various Universities throughout the World. By HERBERT JUNIUS HARDWICKE, M.D., M.R C.P. 8vo. Cloth, $3.00

HARE. Tobacco, Its Physiological and Pathological Effects. 8vo. Illustrated. The Fiske Fund Prize Dissertation for 1885. Paper Covers, .50

HARLAN. Eyesight and How to Care for It. By GEORGE C. HARLAN, M.D., Prof. of Diseases of the Eye, Philadelphia Polyclinic. Illustrated. Cloth, .50

HARLEY. Diseases of the Liver, With or Without Jaundice. Diagnosis and Treatment. By GEORGE HARLEY, M.D. With Colored Plates and Numerous Illustrations. 8vo. Price reduced. Cloth, $3.00 ; Leather, $4.00

HARRIS'S Principles and Practice of Dentistry Including Anatomy, Physiology, Pathology, Therapeutics, Dental Surgery and Mechanism. By CHAPIN A. HARRIS, M.D., D.D.S., late President of the Baltimore Dental College, author of " Dictionary of Medical Terminology and Dental Surgery." Eleventh Edition. Revised and Edited by FERDINAND J. S. GORGAS, A.M., M.D., D.D.S., author of " Dental Medicine ;" Professor of the Principles of Dental Science, Dental Surgery and Dental Mechanism in the University of Maryland. Two Full-page Plates and 744 Illustrations. 994 pages. 8vo. Cloth, $6.50 ; Leather, $7.50

> **Medical and Dental Dictionary.** A Dictionary of Medical Terminology, Dental Surgery, and the Collateral Sciences. Fourth Edition, carefully Revised and Enlarged. By FERDINAND J. S. GORGAS, M.D., D.D.S., Prof. of Dental Surgery in the Baltimore College, 8vo. Cloth, $6.50 ; Leather, $7.50

HARTSHORNE. Our Homes. Their Situation, Construction, Drainage, etc. By HENRY HARTSHORNE, M.D. Illustrated. Cloth, .50

HEADLAND'S Action of Medicines. On the Action of Medicines in the System. By F. W. HEADLAND, M.D. Ninth American Edition. 8vo. Cloth, $3.00

HEATH'S Operative Surgery. A Course of Operative Surgery, consisting of a Series of Plates, Drawn from Nature by M. Léveillé, of Paris. With Descriptive Text of Each Operation. By CHRISTOPHER HEATH, F.R.C.S., Holme Professor of Clinical Surgery in University College, London. Quarto. Second Edition. Revised. *Sold by Subscription.* Cloth, $12.00

> **Minor Surgery and Bandaging.** Seventh Edition, Revised and Enlarged. With 115 Illustrations. 12mo. *Preparing.*

> **Practical Anatomy.** A Manual of Dissections. Sixth London Edition. 24 Colored Plates, and nearly 300 other Illustrations. Cloth, $5.00

> **Injuries and Diseases of the Jaws.** Third Edition. Revised, with over 150 Illustrations. 8vo. Cloth, $4.50

> **Surgical Diagnosis.** Cloth, $1.25

HIGGINS' Ophthalmic Practice. A Handbook for Students and Practitioners. By CHARLES HIGGINS, F.R.C.S. Ophthalmic Assistant Surgeon at Guy's Hospital. Second Edition. 16mo. Cloth, .50

HILLIER. Diseases of Children. A Clinical Treatise. By THOMAS HILLIER, M.D. 8vo. Paper, .75 ; Cloth, $1.25

HILL AND COOPER. Venereal Diseases. The Student's Manual of Venereal Diseases, being a concise description of those Affections and their Treatment. By BERKELY HILL, M.D., Professor of Clinical Surgery, University College, and ARTHUR COOPER, M.D., Late House Surgeon to the Lock Hospital, London. 4th Edition. 12mo. Cloth. *in Press.*

HODGE'S Note-book for Cases of Ovarian Tumors. By H. LENOX HODGE, M.D. With Diagrams. Paper. .50

HODGE on Fœticide or Criminal Abortion. By HUGH L. HODGE, M.D.
Paper, .30; Cloth, .50

HOLDEN'S Anatomy. A Manual of the Dissections of the Human Body. By LUTHER HOLDEN, F.R.C.S. Fifth Edition. Carefully Revised and Enlarged. Specially concerning the Anatomy of the Nervous System, Organs of Special Sense, etc. By JOHN LANGTON, F.R.C.S., Surgeon to, and Lecturer on Anatomy at, St. Bartholomew's Hospital. 208 Illustrations. 8vo.
Oil Cloth Covers, for the Dissecting Room, $4.50; Cloth, $5.00; Leather, $6.00
Landmarks. Medical and Surgical. Third London Edition. Revised and Enlarged. *New Edition Preparing.*
Human Osteology. Comprising a Description of the Bones, with Colored Delineations of the Attachments of the Muscles. The General and Microscopical Structure of Bone and its Development. Carefully Revised. By the Author and A DORAN, F.R.C.S., with Lithographic Plates and Numerous Illustrations. Sixth Edition. 8vo. Cloth, $6.00

HOLDEN. The Sphygmograph. Its Physiological and Pathological Indications. By EDGAR HOLDEN, M.D Illustrated. 8vo. Cloth, $2.00

HORWITZ'S Compend of Surgery, including Minor Surgery, Amputations, Fractures, Dislocations, Surgical Diseases, etc., with Differential Diagnosis and Treatment. By ORVILLE HORWITZ, B.S., M D. Second Edition, Enlarged. 62 Illustrations. 12mo. Cloth, $1.00
Interleaved for the addition of notes, $1.25.

HUFELAND. Long Life. Art of Prolonging Life. By C. W. HUFELAND. Edited by ERASMUS WILSON, M.D. 12mo. Cloth, $1.00

HUGHES. Compend of the Practice of Medicine. Second Edition. Revised and Enlarged. By DANIEL E. HUGHES, M.D., Demonstrator of Clinical Medicine at Jefferson Medical College, Philadelphia. In two parts:
PART I.—Continued, Eruptive and Periodical Fevers, Diseases of the Stomach, Intestines, Peritoneum, Biliary Passages, Liver, Kidneys, etc., and General Diseases, etc.
PART II.—Diseases of the Respiratory System, Circulatory System and Nervous System; Diseases of the Blood, etc.
Price of each Part, in Cloth, $1.00; interleaved for the addition of Notes, $1.25
Physicians' Edition.—In one volume, including a section on Skin Diseases.
Full Morocco, $2.50

HUNTER. Mechanical Dentistry. A Practical Treatise on the Construction of the Various kinds of Artificial Dentures, with Formulæ, Receipts, etc. By CHARLES HUNTER, D.D.S. 100 Illustrations. 12mo. Cloth, $1.50

HUTCHINSON'S Clinical Surgery. Consisting of Plates, Photographs, Woodcuts, Diagrams, etc. Illustrating Surgical Diseases, Symptoms and Accidents; also Operations and other Methods of Treatment. With Descriptive Letter-press. By JONATHAN HUTCHINSON, F.R.C.S., Senior Surgeon to the London Hospital, Surgeon to the Moorfields Ophthalmic Hospital. Imperial 4to. Volume 1. (Ten Parts) bound in cloth, complete in itself, $25.00. Parts Eleven to Seventeen of Volume 2 Now Ready. Each, $2.50

JAMES on Sore Throat. Its Nature, Varieties and Treatment, including its Connection with other Diseases. By PROSSER JAMES, M.D. Fourth Edition, Revised and Enlarged. With Colored Plates and Numerous Wood-cuts. 12mo.
Paper .75; Cloth, $1.25

JONES' Aural Atlas. An Atlas of Diseases of the Membrana Tympani. Being a Series of Colored Plates, containing 62 Figures. With Explanatory Text. By H. MACNAUGHTON JONES, M.D., Surgeon to the Cork Ophthalmic and Aural Hospital. 4to. Cloth, $4.00
Aural Surgery. A Practical Handbook on Aural Surgery. Illustrated. Second Edition, Revised and Enlarged, with new Wood Engravings. 12mo.
Cloth, $2.75

JONES and SIEVEKING'S Pathological Anatomy. A Manual of Pathological Anatomy. By C. HANDFIELD JONES, M.D, and EDWARD H. SIEVEKING, M.D. A New Enlarged Edition. Edited by J. F. PAYNE, M.D. Illus. Cloth, $5.50

JONES. Defects of Sight and Hearing, their Nature, Causes and Prevention. By T. WHARTON JONES, M.D. Second Edition. 16mo. Cloth, .50

KANE'S Drugs that Enslave. The Opium, Morphine, Chloral, and Similar Habits. By H. H. KANE, M.D. With Illustrations. Paper, .75 ; Cloth, $1.25

KIDD'S Laws of Therapeutics; or, the Science and Art of Medicine. By JOSEPH KIDD, M.D. 12mo. Paper, .75; Cloth, $1.25

KIRBY. Selected Remedies. A Pharmacopœia of Selected Remedies, with Therapeutic Annotations, Notes on Alimentation in Disease, Air, Massage, Electricity and other Supplementary Remedial Agents, and a Clinical Index; arranged as a Handbook for Prescribers. By E. A. KIRBY, M.D. Sixth Edition, Enlarged and Revised. 4to. Cloth, $2.25

KIRKE'S Physiology. A Handbook of Physiology. By KIRKE. Eleventh London Edition, Revised and Enlarged. By W. MORRANT BAKER, M.D. 460 Illustrations. Cloth, $4.00; Leather, $5.00

KOLLMYER. Chemia Coartata. A Key to Chemistry. By A. H. KOLLMYER, M.D. With Tables, etc. Cloth, $2.25

LANDIS' Compend of Obstetrics; especially adapted to the Use of Students and Physicians. By HENRY G. LANDIS, M.D., Professor of Obstetrics and Diseases of Women, in Starling Medical College, Columbus, Ohio. Second Edition. Revised and Enlarged. With New Illustrations.
Cloth, $1.00; interleaved for the addition of Notes, $1.25

LANDOIS. A Manual of Human Physiology; including Histology and Microscopical Anatomy, with special reference to the requirements of Practical Medicine. By DR. L. LANDOIS, of the University of Greifswald. Translated from the Fourth German Edition, with additions, by WM. STIRLING, M.D., D.SC., Professor of the Institutes of Medicine in the University of Aberdeen. With very numerous Illustrations. 8vo. 2 Volumes. Cloth, $10.00

LEBER AND ROTTENSTEIN. Dental Caries and Its Causes. An Investigation into the Influence of Fungi in the Destruction of the Teeth. By Drs. LEBER and ROTTENSTEIN. Illustrated. Paper, .75 ; Cloth, $1.25

LEE. The Microtomist's Vade Mecum. By ARTHUR BOLLES LEE. A Handbook of the Methods of Microscopical Anatomy. 660 Formulæ, etc. 12mo.
Cloth, $3.00

LEFFMANN'S Organic and Medical Chemistry. Including Urine Analysis and the Analysis of Water and Food. By HENRY LEFFMANN, M.D., Demonstrator of Chemistry at Jefferson Medical College, Philadelphia. Quiz-Compend Series. 12mo. Cloth, $1.00. Interleaved for the addition of Notes, $1 25

LEGG on the Urine. Practical Guide to the Examination of the Urine, for Practitioner and Student. By J. WICKHAM LEGG, M.D. Sixth Edition, Enlarged. Illustrated. 12mo. Cloth, .75

LEWIN on Syphilis. The Treatment of Syphilis. By Dr. GEORGE LEWIN, of Berlin. Translated by CARL PROEGLER, M.D., and E. H. GALE, M.D., Surgeons U. S. Army. Illustrated. 12mo. Paper, .75 ; Cloth, $1.25

LIEBREICH'S Atlas of Ophthalmoscopy, composed of 12 Chromo-Lithographic Plates (containing 59 Figures), with Text. Translated by H. R. SWANZY, M.D. Third Edition. 4to. Boards. $15.00

LINCOLN. School and Industrial Hygiene. By D. F. LINCOLN, M.D. Cloth, .50

LONGLEY'S Pocket Medical Dictionary for Students and Physicians. Giving the Correct Definition and Pronunciation of all Words and Terms in General Use in Medicine and the Collateral Sciences, with an Appendix, containing Poisons and their Antidotes, Abbreviations Used in Prescriptions, and a Metric Scale of Doses. By ELIAS LONGLEY. Cloth, $1.00; Tucks and Pocket, $1.25

LIZARS. On Tobacco. The Use and Abuse of Tobacco. By JOHN LIZARS, M.D. 12mo. Cloth, .50

MACDONALD'S Microscopical Examinations of Water and Air. A Guide to the Microscopical Examination of Drinking Water, with an Appendix on the Microscopical Examination of Air. By J. D. MACDONALD, M.D. With 25 Lithographic Plates, Reference Tables, etc. Second Ed., Revised. 8vo. Cloth, $2.75

MACKENZIE on the Throat and Nose. Complete. Including the Pharynx, Larynx, Trachea, Œsophagus, Nasal Cavities, etc., etc. By MORELL MACKENZIE, M.D., Senior Physician to the Hospital for Diseases of the Chest and Throat, Lecturer on Diseases of the Throat at London Hospital Medical College, etc.

Vol. I.   Including the Pharynx, Larynx, Trachea, etc.   112 Illustrations.
Vol. II. Including the Œsophagus, Nose, Naso-Pharynx, etc.   Illustrated.
The two Volumes, Cloth, $6.00 ; Leather, $7.50
Volume II, sold separately, Cloth, $3.00 ; Leather, $4.00

Author's Edition, issued under his supervision, containing all the original Wood Engravings, and the essay on " Diphtheria, its Causes, Nature, and Treatment," formerly published separately.

The Pharmacopœia of the Hospital for Diseases of the Throat and Nose. Fourth Edition, Enlarged, Containing 250 Formulæ, with Directions for their Preparation and Use. 16mo.   Cloth, $1.25

Growths in the Larynx. Their History, Causes, Symptoms, etc. With Reports and Analysis of one Hundred Cases. With Colored and other Illustrations. 8vo.   Paper, .75 ; Cloth, $1.25

Hay Fever; Its Etiology and Treatment.   Paper, .50

MAC MUNN. On the Spectroscope in Medicine. By CHAS. A. MAC MUNN, M.D. With 3 Chromo-lithographic Plates of Physiological and Pathological Spectra, and 13 Wood Cuts. 8vo.   Cloth, $3.00

MACNAMARA. On the Eye. A Manual of the Diseases of the Eye. By C. MACNAMARA, M.D. Fourth Edition, Carefully Revised ; with Additions and Numerous Colored Plates, Diagrams of Eye, Wood-cuts, and Test Types. Demi 8vo.   Cloth, $4.00

MADDEN.   Health Resorts for the Treatment of Chronic Diseases. The result of the author's own observations during several years of health travel in many lands. With remarks on climatology and the use of mineral waters. By T. M. MADDEN, M.D. 8vo.   Cloth, $2.50

MANN'S Manual of Psychological Medicine and Allied Nervous Diseases. Their Diagnosis, Pathology, Prognosis and Treatment, including their Medico-Legal Aspects ; with chapter on Expert Testimony, and an abstract of the laws relating to the Insane in all the States of the Union. By EDWARD C. MANN, M.D., member of the New York County Medical Society. With Illustrations of Typical Faces of the Insane, Handwriting of the Insane, and Micro-photographic Sections of the Brain and Spinal Cord. Octavo.   Cloth, $5.00 ; Leather $6.00

MARSHALL.   Anatomical Plates; or Physiological Diagrams. Life Size (4 by 7 feet.) Beautifully Colored. By JOHN MARSHALL, F.R.S. New Edition. Revised and Improved, Illustrating the Whole Human Body.

The Set, 11 Maps, in Sheets, $50.00
The Set, 11 Maps, on Canvas, with Rollers, and Varnished, $80.00
An Explanatory Key to the Diagrams,   .50

No. 1. The Skeleton and Ligaments.   No. 2. The Muscles, Joints, and Animal Mechanics.   No. 3. The Viscera in Position—The Structure of the Lungs.   No. 4. The Organs of Circulation.   No. 5. The Lymphatics or Absorbents.   No. 6. The Digestive Organs.   No. 7. The Brain and Nerves. No. 8. The Organs of the Senses and Organs of the Voice, Plate 1.   No. 9. The Organs of the Senses, Plate 2.   No. 10. The Microscopic Structure of the Textures.   No. 11. The Microscopic Structure of the Textures.

MARSHALL & SMITH. On the Urine. The Chemical Analysis of the Urine. By JOHN MARSHALL, M.D., and Prof. EDGAR F. SMITH, of the Chemical Laboratories, University of Pennsylvania.   Illustrated by Phototype Plates. 12mo.
Cloth, $1.00

MARTIN'S Microscopic Mounting. A Manual. With Notes on the Collection and Examination of Objects. 150 Illustrations. By JOHN H. MARTIN. Second Edition, Enlarged. 8vo.   Cloth, $2.75

MATTHIAS' Legislative Manual. Rules for Conducting Business in Meetings of Societies, Legislative Bodies, Town and Ward Meetings, etc. By BENJ. MATTHIAS, A.M. Eighteenth Edition. Cloth, .50

MAYS' Therapeutic Forces ; or, The Action of Medicine in the Light of the Doctrine of Conservation of Force. By THOMAS J. MAYS, M.D. Cloth, $1.25

MEADOWS' Obstetrics. A Text-Book of Midwifery. Including the Signs and Symptoms of Pregnancy, Obstetric Operations, Diseases of the Puerperal State, etc. By ALFRED MEADOWS, M.D. Third American, from Fourth London Edition. Revised and Enlarged. With 145 Illustrations. 8vo. Cloth, $2.00

MEARS' Practical Surgery. Including : Part I.—Surgical Dressings ; Part II.—Bandaging ; Part III—Ligations ; Part IV.—Amputations. With 490 Illustrations. By J. EWING MEARS, M.D., Demonstrator of Surgery in Jefferson Medical College, and Professor of Anatomy and Clinical Surgery in the Pennsylvania College of Dental Surgery. Second Edition, Revised. 12mo. 794 pages. Cloth, $3.75 ; Sheep, $4.75

MEDICAL Directory of Philadelphia, Pennsylvania, Delaware and Southern half of New Jersey, containing lists of Physicians *of all Schools of Practice*, Dentists, Druggists and Chemists, with information concerning Medical Societies, Colleges and Associations, Hospitals, Asylums, Charities, etc. Published Annually. For 1885 now Ready. 12mo. Full Morocco, Gilt edges, $2.50

MEIGS. Milk Analysis and Infant Feeding. A Practical Treatise on the Examination of Human and Cows' Milk, Cream, Condensed Milk, etc., and Directions as to the Diet of Young Infants. By ARTHUR V. MEIGS, M.D., Physician to the Pennsylvania Hospital, Philadelphia. 12mo. Cloth, $1.00

MEIGS and PEPPER on Children. A Practical Treatise on the Diseases of Children. By J. FORSYTH MEIGS, M.D., Fellow of the College of Physicians of Philadelphia, etc., etc., and WILLIAM PEPPER, M.D., Professor of the Principles and Practice of Medicine in the Medical Department, University of Pennsylvania. Seventh Edition. Cloth, $6.00 ; Leather, $7.00

MENDENHALL'S Vade Mecum. The Medical Student's Vade Mecum. A Compend of Anatomy, Physiology, Chemistry, The Practice of Medicine, Surgery, Obstetrics, etc. By GEO. MENDENHALL, M.D. 11th Ed. 224 Illus. 8vo.

MERRELL'S Digest of Materia Medica. Forming a Complete Pharmacopœia for the use of Physicians, Pharmacists and Students. By ALBERT MERRELL, M.D. Octavo. Half dark Calf, $4.00

MILLER and LIZAR'S Alcohol and Tobacco. Alcohol. Its Place and Power. By JAMES MILLER, F.R.C.S. ; and, Tobacco, Its Use and Abuse. By JOHN LIZARS, M.D. The two essays in one volume. Cloth, $1.00 ; Separate, each .50

MORRIS on the Joints. The Anatomy of the Joints of Man. Comprising a Description of the Ligaments, Cartilages and Synovial Membranes ; of the Articular Parts of Bones, etc. By HENRY MORRIS, F.R.C.S. Illustrated by 44 Large Plates and Numerous Figures, many of which are Colored. 8vo. Cloth, $5.50

MORTON on Refraction of the Eye. Its Diagnosis and the Correction of its Errors. With Chapter on Keratoscopy. By A. MORTON, M.B. Second Ed. Cloth, $1.00

MUTER'S Chemistry. An Introduction to Pharmaceutical and Medical Chemistry. Part I.—Theoretical and Descriptive. Part II.—Practical and Analytical. By JOHN MUTER, M.D., President of the Society of Public Analysts. Second Edition, Enlarged and Rearranged. The Two Parts in one volume. 8vo. Cloth, $6.00

Part II. Practical and Analytical. Bound Separately. Cloth, $2.50

OSGOOD. The Winter and Its Dangers. By HAMILTON OSGOOD, M.D. Cloth, .50

OTT'S Action of Medicines. By ISAAC OTT, M.D., late Demonstrator of Experimental Physiology in the University of Pennsylvania. 22 Illus. Cloth, $2.00

OVERMAN'S Practical Mineralogy, Assaying and Mining, with a Description of the Useful Minerals, etc. By FREDERICK OVERMAN, Mining Engineer. Eleventh Edition. 12mo. Cloth, $1.00

OLDBERG'S Three Hundred New Prescriptions. Selected Chiefly from the Best Collections of Formulæ used in Hospital and Out-patient practice, with a Dose Table, and a Complete Account of the Metric System. By OSCAR OLDBERG, PHAR. D., Member of the Pharmaceutical Association, and of the Sixth Decennial Committee of Revision and Publication of the Pharmacopœia of the United States. Paper Covers, .75; Cloth, $1.25

The Unofficial Pharmacopœia. Comprising over 700 Popular and Useful Preparations, not Official in the United States, of the various Elixirs, Fluid Extracts, Mixtures, etc., in constant demand. Half Leather, $3.50

PACKARD'S Sea Air and Sea Bathing. By JOHN H. PACKARD, one of the Physicians to the Pennsylvania Hospital, Philadelphia. Cloth, .50

PAGE'S Injuries of the Spine and Spinal Cord, without apparent Lesion and Nervous Shock. In their Surgical and Medico-Legal Aspects. By HERBERT W. PAGE, M.D., F.R.C.S. Second Edition, Revised. Octavo. Cloth, $3.50

PAGET'S Lectures on Surgical Pathology. Delivered at the Royal College of Surgeons. By JAMES PAGET, F.R.S. Third Edition. Edited by WILLIAM TURNER, M.D. With Numerous Illustrations. 8vo. Cloth, $7.00; Leather, $8.00

PARKES' Practical Hygiene. By EDWARD A. PARKES, M.D. The Sixth Revised and Enlarged Edition. With Many Illustrations. 8vo. Cloth, $3.00

PARRISH'S Alcoholic Inebriety. From a Medical Standpoint, with Illustrative Cases from the Clinical Records of the Author. By JOSEPH PARRISH, M.D., President of the Amer. Assoc. for Cure of Inebriates. Paper, .75; Cloth, $1.25

PENNSYLVANIA Hospital Reports. Edited by a Committee of the Hospital Staff: J. M. DACOSTA, M.D., and WILLIAM HUNT. Containing Original Articles by the Staff. With many other Illustrations. Paper, .75; Cloth, $1.25

PEREIRA'S Prescription Book. Containing Lists of Terms, Phrases, Contractions and Abbreviations used in Prescriptions, Explanatory Notes, Grammatical Construction of Prescriptions, Rules for the Pronunciation of Pharmaceutical Terms. By JONATHAN PEREIRA, M.D. Sixteenth Edition.
Cloth, $1.00; Leather, with tucks and pocket, $1.25

PHYSICIAN'S VISITING LIST. Published Annually. Thirty-fourth Year of its Publication.

SIZES AND PRICES.

For 25 Patients weekly.		Tucks, pocket and pencil,			$1.00
50	" "	" " "			1.25
75	" "	" " "			1.50
100	" "	" " "			2.00
50	" " 2 vols.	{ Jan. to June } { July to Dec. }		"	2.50
100	" " 2 vols.	{ Jan. to June } { July to Dec. }		"	3.00

INTERLEAVED EDITION.

For 25 Patients weekly, interleaved, tucks, pocket, etc.,					1.25
50	" "	" " " "			1.50
50	" " 2 vols.	{ Jan. to June } { July to Dec. }	"	"	3.00

Perpetual Edition, without Dates and with Special Memorandum Pages.
For 25 Patients, interleaved, tucks, pocket and pencil, ... $1.25
50 " " " " " " 1.50
EXTRA Pencils will be sent, postpaid, for 25 cents per half dozen.

PIESSE'S Art of Perfumery, or the Methods of Obtaining the Odors of Plants, and Instruction for the Manufacture of Perfumery, Dentrifices, Soap, Scented Powders, Cosmetics, etc. By G. W. SEPTIMUS PIESSE. Fourth Edition. 366 Illustrations. Cloth, $5.50

PIERSOL. Normal Histology. A Synopsis. Adapted to the course at the University of Penn'a. By GEORGE A. PIERSOL, M.D., Demonstrator of Histology. 40 Photoand Micrographic Plates, containing 200 Figures. $6.50

POTTER. A Handbook of Materia Medica, Pharmacy and Therapeutics, including the Action of Medicines, Special Therapeutics, Pharmacology, etc. By SAMUEL O. L. POTTER, M.A., M.D.   *Nearly Ready.*

    Speech and Its Defects. Considered Physiologically, Pathologically and Remedially; being the Lea Prize Thesis of Jefferson Medical College, 1882. Revised and Corrected. 12mo.   Cloth, $1.00

    Compend of Anatomy. 63 Illustrations. Third Edition, Revised.

    Compend of Visceral Anatomy. Illustrated. Second Edition.

    Compend of Materia Medica and Therapeutics, arranged in accordance with the Sixth Revision U. S. Pharmacopœia. Revised Edition, with Index.
        Price for each, Cloth, $1.00; Interleaved for taking Notes, $1.25

PIGGOTT on Copper Mining and Copper Ore. With a full Description of the Principal Copper Mines of the United States, the Art of Mining, etc. By A. SNOWDEN PIGGOTT. 12mo.   Cloth, $1.00

POWER, HOLMES, ANSTIE and BARNES (Drs.) Reports on the Progress of Medicine, Surgery, Physiology, Midwifery, Diseases of Women and Children, Materia Medica, Medical Jurisprudence, Ophthalmology, etc. Reported for the New Syndenham Society. 8vo.   Paper, .75; Cloth, $1.25

PRINCE'S Plastic and Orthopedic Surgery. By DAVID PRINCE, M.D. Containing a Report on the Condition of, and Advance made in, Plastic and Orthopedic Surgery, etc. Numerous Illustrations. 8vo.   Cloth, $4.50

PROCTER'S Practical Pharmacy. Lectures on Practical Pharmacy. With 43 Engravings and 32 Lithographic Fac-simile Prescriptions. By BARNARD S. PROCTER. Second Edition.   Cloth, $4.50

PYE. Surgical Handicraft. A Manual of Surgical Manipulations, Minor Surgery and other Matters connected with the work of Surgeons, Surgeons' Assistants, etc. By WALTER PYE, M.D., Surgeon to St. Mary's Hospital, London. 208 Illustrations.   Cloth, $5.00

RADCLIFFE on Epilepsy, Pain, Paralysis, and other Disorders of the Nervous System. By CHARLES BLAND RADCLIFFE, M.D. Illus. Paper, .75; Cloth, $1.25

RALFE. Diseases of the Kidney and Urinary Derangements. By C. H. RALFE, M.D., F.R.C.P., Ass't Physician to the London Hospital. Illustrated. 12mo. Volume 3. *Practical Series.*   Cloth, $2.75

RECORD for the Sick Room. Designed for the Use of Nurses and others engaged in caring for the Sick. It consists of Blanks, in which may be recorded the Hour, State of Pulse, Temperature, Respiration, Medicines to be Given, Food Taken, etc., together with a List of Directions for the Nurse to pursue in Emergencies. *Sample Pages Free.*   One Copy, .25; Per Dozen, $2.50

REESE'S Medical Jurisprudence and Toxicology. A Text-book for Medical and Legal Practitioners and Students. By JOHN J. REESE, M.D., Editor of Taylor's Jurisprudence, Professor of the Principles and Practice of Medical Jurisprudence, including Toxicology, in the University of Pennsylvania Medical and Law Schools. Crown Octavo.   Cloth, $4.00; Leather, $5.00

REEVES. Bodily Deformities and their Treatment. A Handbook of Practical Orthopædics. By H. A. REEVES, M.D., Senior Ass't Surgeon to the London Hospital, Surgeon to the Royal Orthopædic Hospital. 228 Illus.   Cloth, $2.25

REYNOLDS. Electricity. Lectures on the Clinical Uses of Electricity. By J. RUSSELL REYNOLDS, M.D., F.R.S. Second Edition. 12mo.   Cloth, $1.00

RICHARDSON. Long Life, and How to Reach It. By J. G. RICHARDSON, Prof. of Hygiene, University of Penna.   Cloth, .50

RICHARDSON'S Mechanical Dentistry. A Practical Treatise on Mechanical Dentistry. By JOSEPH RICHARDSON, D.D.S. Fourth Edition. With 185 Illustrations. 8vo.   Cloth, $4.00; Leather, $4.75

RIGBY'S Obstetric Memoranda. Fourth Edition, Revised. By ALFRED MEADOWS, M.D. 32mo.   Cloth, .50

RICHTER'S Inorganic Chemistry. A Text-book for Students. By Prof. VICTOR VON RICHTER, University of Breslau. Second American, from Fourth German Edition. Authorized Translation by EDGAR F. SMITH, M.A., PH.D., Prof. of Chemistry, Wittenberg College, formerly in the Laboratories of the University of Pennsylvania, Member of the Chemical Societies of Berlin and Paris. With 89 Illustrations and a Colored Plate of Spectra. 12mo. 424 pages. Cloth, $2.00

Organic Chemistry. A Text-book for Students. Authorized translation from the Fourth German Edition, by Prof. Edgar F. Smith. Illus. Cloth, $3.00

RINDFLEISCH'S General Pathology. A Handbook for Students and Physicians. By Prof. EDWARD RINDFLEISCH, of Wurzburg. Translated by WM. H. MERCUR, M.D., of Pittsburgh, Pa., Edited and Revised by JAMES TYSON, M.D., Professor of Morbid Anatomy and Pathology, University of Pennsylvania. Cloth, $2.00

ROBERTS. Practice of Medicine. The Theory and Practice of Medicine. By FREDERICK ROBERTS, M.D., Professor of Therapeutics at University College, London. Fifth American Edition, thoroughly revised and enlarged, with New Illustrations. 8vo. Cloth, $5.00; Leather, $6.00

Recommended as a text-book at the University of Pennsylvania, Yale and Dartmouth Colleges, University of Michigan, and many other Medical Schools.

Materia Medica and Pharmacy. A Compend for Students. 12mo.
Cloth, $2.00

ROBERTS. Surgical Delusions and Follies. By JOHN B. ROBERTS, M.D., Professor of Anatomy and Surgery, in the Philadelphia Polyclinic. Paper, .25; Cloth, .50

The Human Brain. The Field and Limitation of the Operative Surgery of the Human Brain. Illustrated. 8vo. Cloth, $1.25

RYAN'S Philosophy of Marriage, in Its Social, Moral and Physical Relations, and Diseases of the Urinary Organs. By MICHAEL RYAN, M.D. Member of the Royal College of Physicians, London. 12mo. Cloth, $1.00

SANDERSON'S Physiological Laboratory. A Handbook of the Physiological Laboratory. Being Practical Exercises for Students in Physiology and Histology. By J. BURDON SANDERSON, M.D., E. KLEIN, M.D., MICHAEL FOSTER, M.D., F.R.S., and T. LAUDER BRUNTON, M.D. With over 350 Illustrations and Appropriate Letter-press Explanations and References. One Volume. Cloth, $5.00
Adopted as a Text-book at Yale College and other Medical Schools.

SANSOM'S Diseases of the Heart. Valvular Disease of the Heart. By ARTHUR ERNEST SANSOM, M.D. Illustrated. 12mo. Cloth, $1.25

On Chloroform. Its Action and Administration. Paper, .75; Cloth, $1.25

SAVAGE. On the Pelvic Organs. The Surgery, Surgical Pathology and Surgical Anatomy of the Female Pelvic Organs. In a Series of Colored Plates taken from Nature, with Commentaries, Notes and Cases. By HENRY SAVAGE, M.D., F.R.C.S. New Edition. Issued by arrangement with the Author, from the original Plates. 4to. Cloth, $12.00

SCHULTZE'S Obstetrical Plates. Obstetrical Diagrams. Life Size. By Prof. B. S. SCHULTZE, M.D., of Berlin. Twenty in the Set. Colored.
In Sheets, $15.00; Mounted on Rollers, $25.00

SIEVEKING on Life Assurance. The Medical Adviser in Life Assurance. By E. H. SIEVEKING, M.D. Second Edition. Revised. 12mo Cloth, $2.00

SOLLY'S Colorado Springs and Manitou as Health Resorts. By S. EDWIN SOLLY, M.D., M.R.C.S., Eng. 12mo. Paper cover, .25

SMITH on Ringworm. Its Diagnosis and Treatment. By ALDER SMITH, F.R.C.S. With Illustrations. 12mo. Cloth, $1.00

SMITH'S Wasting Diseases of Infants and Children. By EUSTACE SMITH, M.D., F.R.C.P., Physician to the East London Children's Hospital. Fourth London Edition, Enlarged. 8vo. Cloth, $3.00

SMYTHE'S Medical Heresies. Historically Considered. A Series of Critical Essays on the Origin and Evolution of Sectarian Medicine, embracing a Special Sketch and Review of Homœopathy, Past and Present. By GONZALVO C. SMYTHE, A.M., M.D., Professor of the Principles and Practice of Medicine, College of Physicians and Surgeons, Indianapolis, Indiana. 12mo. Cloth, $1.25

STAMMER. Chemical Problems, with Explanations and Answers. By KARL STAMMER. Translated from the 2d German Edition, by Prof. W. S. HOSKINSON, A.M., Wittenberg College, Springfield, Ohio. 12mo. Cloth.   .75

STARR. The Digestive System in Childhood. The Diseases of the Digestive System in Infancy and Childhood. By LOUIS STARR, M.D., Clinical Professor of Diseases of Children in the Hospital of the University of Pennsylvania; Physician to the Children's Hospital, Philadelphia.   *In Press.*

STEWART'S Compend of Pharmacy. By F. E. STEWART, MD., PH.G., Quiz Master in Chemistry and Theoretical Pharmacy, Philadelphia College of Pharmacy; Demonstrator and Lecturer in Pharmacology, Medico-Chirurgical College, etc., etc. Quiz Compend Series.
          Cloth, $1.00; Interleaved for the addition of notes, $1.25

STOCKEN'S Dental Materia Medica. The Elements of Dental Materia Medica and Therapeutics with Pharmacopœia. By JAMES STOCKEN, D.D.S. Third Edition. 12mo.   Cloth, $2.50

SUTTON'S Volumetric Analysis. A Systematic Handbook for the Quantitative Estimation of Chemical Substances by Measure, Applied to Liquids, Solids and Gases. By FRANCIS SUTTON, F.C.S. Fourth Edition, Revised and Enlarged, with Illustrations. 8vo.   Cloth, $5.00

SWAYNE'S Obstetric Aphorisms, for the Use of Students commencing Midwifery Practice. By JOSEPH G. SWAYNE, M.D. Eighth Edition. Illus.   Cloth, $1.25

SWERINGEN'S Druggists' Reference Book. A Pharmaceutical Lexicon or Dictionary of Pharmaceutical Science. Containing Explanations of the various Subjects and Terms of Pharmacy, with appropriate Selections from the Collateral Sciences. Formulæ for Officinal, Empirical and Dietetic Preparations, etc. By HIRAM V. SWERINGEN, M.D. 8vo.   Cloth, $3.00; Leather, $4.00

TAFT'S Operative Dentistry. A Practical Treatise on Operative Dentistry. By JONATHAN TAFT, D.D.S. Fourth Revised and Enlarged Edition. Over 100 Illustrations. 8vo.   Cloth, $4.25; Leather, $5.00

     Index of Dental Periodical Literature. 8vo.   *In Press.*

TANNER'S Index of Diseases and their Treatment. By THOS. HAWKES TANNER, M.D., F.R.C.P. Second Edition, Revised and Enlarged. By W. H. BROADBENT, M.D. With Additions. Appendix of Formulæ, etc. 8vo.   Cloth, $3.00

     Memoranda of Poisons and their Antidotes and Tests. Fifth American, from the Last London Edition. Revised and Enlarged.   Cloth, .75

TEMPERATURE Charts for Recording Temperature, Respiration, Pulse, Day of Disease, Date, Age, Sex, Occupation, Name, etc.   Put up in pads; each .50

TILT'S Change of Life in Women, in Health and Disease. A Practical Treatise on the Diseases incidental to Women at the Decline of Life. By EDWARD JOHN TILT, M.D. Fourth London Edition. 8vo.   Paper cover, .75; Cloth, $1.50

THOMPSON. Lithotomy and Lithotrity. Practical Lithotomy and Lithotrity; or, an Inquiry into the best Modes of Removing Stone from the Bladder. By Sir HENRY THOMPSON, F.R.C.S., Emeritus Professor of Clinical Surgery in University College. Third Edition. With 87 Engravings. 8vo.   Cloth, $3.50

     Urinary Organs. Diseases of the Urinary Organs. Clinical Lectures. Seventh London Edition, Enlarged, with 73 Illustrations.   Cloth, $1.25

     On the Prostate. Diseases of the Prostate. Their Pathology and Treatment. Fifth London Edition. 8vo. Illustrated.   Cloth, $1.25

     Calculous Diseases. The Preventive Treatment of Calculous Disease, and the Use of Solvent Remedies. Second Edition. 16mo.   Cloth, $1.00

     Surgery of the Urinary Organs. Lectures on some Important Points connected with the Surgery of the Urinary Organs. Illustrated. 8vo.
          Paper, .75; Cloth, $1.25

     Tumors of the Bladder. Their Nature, Symptoms and Surgical Treatment. Preceded by a Consideration of the Best Methods of Diagnosing all Forms of Vesical Disease. Illustrated. 8vo.   Cloth, $1.75

     Stricture of the Urethra and Urinary Fistulæ, their Pathology and Treatment. Fourth Edition. Illustrated.   Cloth, $2.00

THOMPSON'S Manual of Physics. A Student's Manual. By SYLVANUS P. THOMPSON, B.A., D.SC., F.R.A.S., Professor of Experimental Physics in University College, Bristol, England. *Preparing.*

THOROWGOOD on Asthma. Its Forms, Nature and Treatment. By JOHN C. THOROWGOOD, M.D. Second Edition. Cloth, $1.50

TOMES' Dental Anatomy. A Manual of Dental Anatomy, Human and Comparative. By C. S. TOMES, D.D.S. 191 Illustrations. 2d Ed. 12mo. Cloth, $4.25

TOMES. Dental Surgery. A System of Dental Surgery. By JOHN TOMES, F.R.S. Fourth Edition, Revised and Enlarged. By C. S. TOMES, D.D.S. With 263 Illustrations. 12mo. *Preparing.*

TRANSACTIONS of the College of Physicians of Philadelphia. New Series. Vols. I, II, III, IV, V, Cloth, each, $2.50. VI, VII, Cloth, each, $3.50

TRANSACTIONS American Surgical Association. Volumes I and II. Illustrated. Edited by J. EWING MEARS, M.D., Recorder of the Association. Royal 8vo. Price of Vol. I, Cloth, $3.50; Vol. II, Cloth, $4.00; Vol. III, Cloth, $3.50

TRIMBLE. Practical and Analytical Chemistry. Being a complete course in Chemical Analysis. By HENRY TRIMBLE, PH.G., Professor of Analytical Chemistry in the Philadelphia College of Pharmacy. Illustrated. 8vo. Cloth, $1.50

TUKE on Sleep Walking and Hypnotism. By D. HACK TUKE, M.D., LL.D., F.R.C.P., Co-Editor of the Journal of Mental Diseases. 8vo. Cloth, $1.75

    History of the Insane in the British Islands. Cloth, $3.50

TURNBULL'S Artificial Anæsthesia. The Advantages and Accidents of Artificial Anæsthesia; Its Employment in the Treatment of Disease; Modes of Administration; Considering their Relative Risks; Tests of Purity; Treatment of Asphyxia; Spasms of the Glottis; Syncope, etc. By LAURENCE TURNBULL, M.D., PH. G., Aural Surgeon to Jefferson College Hospital, etc. Second Edition, Revised and Enlarged. With 27 Illustrations of Various Forms of Inhalers, etc., and an appendix of over 70 pages, containing a full account of the new local Anæsthetic, Hydrochlorate of Cocaine. 12mo. Cloth, $1.50

    Hydrochlorate of Cocaine. 12mo. Paper, .50

TUSON. Veterinary Pharmacopœia. Including the Outlines of Materia Medica and Therapeutics. For the Use of Students and Practitioners of Veterinary Medicine. By RICHARD V. TUSON, F.C.S. Third Edition. 12mo. Cloth, $2.50

TYSON. Bright's Disease and Diabetes. With Especial Reference to Pathology and Therapeutics. By JAMES TYSON, M.D., Professor of Pathology and Morbid Anatomy in the University of Pennsylvania. With Colored Plates and many Wood Engravings. 8vo. Cloth, 3.50

    Guide to the Examination of Urine. Fifth Edition. For the Use of Physicians and Students. With Colored Plates and Numerous Illustrations Engraved on Wood. Fifth Edition. Enlarged and Revised. 12mo. 249 pages. Cloth, $1.50

    Cell Doctrine. Its History and Present State. With a Copious Bibliography of the subject. Illustrated by a Colored Plate and Wood Cuts. Second Edition. 8vo. Cloth, $2.00

    Rindfleisch's Pathology. Edited by Prof. TYSON. General Pathology; a Handbook for Students and Physicians. By Prof. EDWARD RINDFLEISCH, of Wurzburg. Translated by WM. H. MERCUR, M.D. Edited and Revised by JAMES TYSON, M.D., Professor of Morbid Anatomy and Pathology, University of Pennsylvania. Cloth, $2.00

VACHER'S Primer of Chemistry. With Analysis. By ARTHUR VACHER. Cl., .50

VALENTIN'S Qualitative Analysis. A Course of Qualitative Chemical Analysis. By WM. G. VALENTIN, F.C.S. Sixth Edition. Revised and Corrected by W. R. HODGKINSON, PH.D. (Würzburg), Fellow of the Institute of Chemistry, and of the Chemical, Physical and Geological Societies of London; Lecturer on Chemistry in the South Kensington Science Schools. Assisted by H. M. CHAPMAN, Assistant Demonstrator of Chemistry in the Royal School of Mines. Illustrated. Octavo. Cloth, $3.00

VIRCHOW'S Post-mortem Examinations. A Description and Explanation of the Method of Performing them in the Dead House of the Berlin Charité Hospital, with especial reference to Medico-legal Practice. By Prof. VIRCHOW. Translated by Dr. T. P. SMITH. Third Edition, with Additions and New Plates. 12mo. Cloth, $1.00

VAN HARLINGEN on Skin Diseases. A Practical Manual of Diagnosis and Treatment. For Students and Practitioners. By ARTHUR VAN HARLINGEN, M.D., Professor of Diseases of the Skin in the Philadelphia Polyclinic; Consulting Physician to the Philadelphia Dispensary for Skin Diseases. Including Formulæ. Illustrated by two Colored Plates, with a number of figures showing the appearance of various lesions. 12mo. Cloth, $1.75

WALKER on Intermarriage; or, the Mode in which, and the Causes why, Beauty, Health and Intellect result from certain Unions; and Deformity, Disease and Insanity from others. Illustrated. 12mo. Cloth, $1.00

WARD'S Compend of Chemistry for Chemical and Medical Students. By G. MASON WARD, M.D., Demonstrator of Chemistry in Jefferson Medical College, Philadelphia. Containing a Table of Elements and Tables for the Detection of Metals in Solutions of Mixed Substances, etc. 12mo. 2d Edition. Cloth, $1.00
Interleaved for the addition of Notes, $1.25

WARING. Practical Therapeutics. A Manual for Physicians and Students. By Edward J. Waring, M.D. Fourth Edition. Revised, Rewritten and Rearranged by DUDLEY W. BUXTON, M.D., Assistant to the Professor of Medicine, University College, London. Crown Octavo. *In Press.*

WARNER. Case Taking. A Manual of Clinical Medicine and Case Taking. By FRANCIS WARNER, M.D. Second Edition. Cloth, $1.75

WATSON on Amputations of the Extremities and Their Complications. By B. A. WATSON, A.M., M.D., Surgeon to the Jersey City Charity Hospital and to Christ's Hospital, Jersey City, N. J.; Member of the American Surgical Association. With over 250 Wood Engravings and two Full-page Colored Plates. Octavo. 770 pages. Cloth, $5.50

WATSON'S Physician's Ledger and Cash Book. Based upon and Designed to be Used with Lindsay and Blakiston's Physician's Visiting List (see page 17). *Sample Pages Free.* Price, for 1000 accounts, Leather, $6.50;
500 accounts, Leather, $5.00; 500 accounts, Cloth, $4.00

WATTS' Chemistry. A Manual of Chemistry, Physical and Inorganic. (Being the 13th Edition of Fowne's Physical and Inorganic Chemistry.) By HENRY WATTS, B.A., F.R.S., Editor of the Journal of the Chemical Society; Author of "A Dictionary of Chemistry," etc. With Colored Plate of Spectra and other Illustra-12mo. 595 pages. Cloth, $2.25

WELCH'S Enteric Fever. Its Prevalence and Modifications; Ætiology, Pathology and Treatment. By FRANCIS H. WELCH, F.R.C.S. 8vo. Cloth, $2.00

WELLS. Abdominal Tumors. Their Diagnosis and Surgical Treatment. By T. SPENCER WELLS, M.D., F.R.C.S., Consulting Surgeon to the Samaritan Hospital for Women, etc. Illustrated. 8vo. Cloth, $1.50
   Ovarian and Uterine Tumors. Their Diagnosis and Surgical Treatment. Illustrated. 8vo.

WEST'S How to Examine the Chest. A Practical Guide for the Use of Students. By SAMUEL WEST, M.D. Oxon., M.R.C.P., Physician to the City of London Hospital for Diseases of the Chest. Illustrated. 32mo. Cloth, $1.75

WHITE. The Mouth and Teeth. By J. W. WHITE, M.D., D.D.S. Editor of the Dental Cosmos. Illustrated. Cloth, .50

WICKES' Sepulture. Its History, Methods and Sanitary Requisites. By STEPHEN WICKES, A.M., M.D. Octavo. Cloth, $1.50

WILKES' Pathological Anatomy. Lectures on Pathological Anatomy. By SAMUEL WILKES, F.R.S. Second Edition, Revised and Enlarged by WALTER MOXON, M.D., F.R.S., Physician to and Lecturer at Guy's Hospital, London. Cloth, $6.00
   The Nervous System. Lectures on Diseases of the Nervous System, Delivered at Guy's Hospital, London. New Edition. 8vo. Cloth, $6.00

WILSON. **The Summer** and Its Diseases. By JAMES C. WILSON, M.D.   Cloth, .50

WILSON'S Drainage for Health; or, Easy Lessons in Sanitary Science, with Illustrations. By JOSEPH WILSON, M.D., Medical Director U. S. N.   Cloth, $1.00

Naval Hygiene; or, Human Health and Means for Preventing Disease. With Illustrative Incidents from Naval Experience. Illus.   Cloth, $3.00

WILSON'S Text-Book of Domestic Hygiene and Sanitary Information. A Guide to Personal and Domestic Hygiene. By GEORGE WILSON, M.D., Medical Officer of Health. Edited by Jos. G. RICHARDSON, M.D., Professor of Hygiene at the University of Pennsylvania.   Cloth, $1.00

Handbook of Hygiene and Sanitary Science. With Illustrations. Fifth Edition, Revised and Enlarged. 8vo.   Cloth, $2.75

WILSON. **Human Anatomy.** The Anatomist's Vade-mecum. General and Special. By Prof. ERASMUS WILSON. Edited by GEORGE BUCHANAN, Professor of Clinical Surgery in the University of Glasgow; and HENRY E. CLARK, Lecturer on Anatomy at the Royal Infirmary School of Medicine, Glasgow. Tenth Edition. With 450 Engravings (including 26 Colored Plates).   Cloth, $6.00

Healthy Skin and Hair. A Practical Treatise. Their Preservation and Management. Eighth Edition. 12mo.   Paper, $1.00

WILSON'S The Ocean as a Health Resort. A Handbook of Practical Information as to Sea Voyages, for the Use of Tourists and Invalids. By WM. S. WILSON, M.D. With a Chart showing the Ocean Routes, and Illustrating the Physical Geography of the Sea.   Cloth, $2.50

WOAKES. **Post-Nasal Catarrh** and Diseases of the Nose, causing Deafness. By EDWARD WOAKES, M.D., Senior Aural Surgeon to the London Hospital for Diseases of the Throat and Chest. 26 Illustrations.   Cloth, $1.50

On Deafness, Giddiness and Noises in the Head, or the Naso-Pharyngeal aspect of Ear Disease. Third Edition. Illustrated.   *In Press.*

WOLFE on the Eye. A Practical Treatise on Diseases and Injuries of the Eye. By M. WOLFE, M.D., Senior Surgeon to the Glasgow Ophthalmic Institution, etc. With 10 Colored Plates, and Numerous other Illustrations. Octavo. Cloth, $7.00

WOLFF. **Manual of Applied Medical Chemistry** for Students and Practitioners of Medicine. By LAWRENCE WOLFF, M.D., Demonstrator of Chemistry in Jefferson Medical College, Philadelphia.   Cloth, $1.50

WOOD. **Brain Work and Overwork.** By Prof. H. C. WOOD, Clinical Professor of Nervous Diseases, University of Pennsylvania. 12mo.   Cloth, .50

WOODMAN and TIDY. **Medical Jurisprudence.** Forensic Medicine and Toxicology. By W. BATHURST WOODMAN, M.D., Physician to the London Hospital, and CHARLES MEYMOTT TIDY, F.C.S., Professor of Chemistry and Medical Jurisprudence at the London Hospital. With Chromo-lithographic Plates, representing the Appearance of the Stomach in Poisoning by Arsenic, Corrosive Sublimate, Nitric Acid, Oxalic Acid; the Spectra of Blood and the Microscopic Appearance of Human and other Hairs; and 116 other Illustrations. Large Octavo. *Sold by Subscription.*   Cloth, $7.50; Leather, $8.50

WRIGHT on Headaches; their Causes, Nature and Treatment. By HENRY G. WRIGHT, M.D. 12mo. *Ninth Thousand.*   Cloth, .50

WYTHE on the Microscope. A Manual of Microscopy and Compendium of the Microscopic Sciences, Micro-mineralogy, Biology, Histology and Practical Medicine, with Index and Glossary and the genera of microscopic plants. By JOSEPH H. WYTHE, A.M., M.D. Fourth Edition. 252 Illus. Cloth, $3.00; Leather, $4.00

Dose and Symptom Book. The Physician's Pocket Dose and Symptom Book. Containing the Doses and Uses of all the Principal Articles of the Materia Medica, and Original Preparations. Sixteenth Revised Edition.
Cloth, $1.00; Leather, with Tucks and Pocket, $1.25

YEO'S Manual of Physiology. A Text-book for Students of Medicine. By GERALD F. YEO, M.D., F.R.C.S., Professor of Physiology in King's College, London. With over 300 carefully printed Illustrations. A Glossary and Complete Index. Crown Octavo.   Cloth, $4.00; Leather, $5.00

A Text-Book of Medical Chemistry.

BY E. H. BARTLEY, M.D.,

Associate Professor of Chemistry at the Long Island College Hospital; President of the American Society of Public Analysts; Chief Chemist, Board of Health, of Brooklyn, N. Y., etc.

Illustrated 12mo. Cloth, $2.50.

This book, written especially for students and physicians, aims to be a text-book for the one and a work of reference for the other. It is practical and concise, dealing only with those parts of chemistry pertaining to medicine; no time being wasted in long descriptions of substances and theories of interest only to the advanced chemical student.

PART I—Treats of Light, Heat and Electricity, which are described at some length, and explanations made and applied to common phenomena. In the subject of light, only so much is given as will explain the theory and use of the spectroscope. In electricity, the principal aim has been to give such information as is needed for the proper understanding, working and care of th~ medical battery.

PART II—Theoretical Chemistry. Only such portions of the well established principles of modern chemistry as are necessary to an understanding of the subject are given. It has been deemed best to present all these elementary parts first, that the student may be better able to study any set of isolated facts. These theories are presented in a concise, clear way, in a logical order and in a manner which the author has found specially successful in an experience of over twelve years of teaching.

PART III—Treats of the natural history of the elements, of their principal compounds, with their physiological action and toxicology.

PART IV—Organic bodies commonly used in medicine and pharmacy. The principal organic substances derived from animal life are given a place. In the appendix will be found analyses of the principal secretions and tissues, tables of solubilities and of specific gravities, the metric system, and other useful information.

Applied Medical Chemistry.

Containing a description of the apparatus and methods employed in the practice of Medical Chemistry, the Chemistry of Poisons, Physiological and Pathological Analysis, Urinary and Fecal Analysis, Sanitary Chemistry and the Examination of Medicinal Agents, Foods, etc.

BY LAWRENCE WOLFF, M.D.,

Demonstrator of Chemistry in the Jefferson Medical College; Member of the Philadelphia College of Pharmacy and of the Chemical Section of the Franklin Institute, etc.

Octavo, Cloth, $1 50.

*** The object of the author of this book is to furnish the practitioner and student a reliable and simple guide for making analyses and examinations of the various medicinal agents, human excretions, secretions, etc., without elaborate apparatus or expensive processes.

Practical and Analytical Chemistry.

Being a complete course in Chemical Analysis, for pharmaceutical and medical students.

BY HENRY TRIMBLE, Ph.G.,

Professor of Analytical Chemistry in the Philadelphia College of Pharmacy.

Illustrated. 8vo. Cloth, $1.50.

SUMMARY OF CONTENTS. Part I. Practical—Preparation and Properties of Gases, Preparation of Salts, etc. Part II. Section I—Bases. Group I—Potassium, Sodium, Lithium, Ammonium. Group II—Barium, Strontium, Calcium, Magnesium. Group III—Manganese, Zinc, Cobalt, Nickel. Group IV—Iron, Cerium, Chromium, Aluminium. Group V—Arsenic, Antimony, Tin, Gold, Platinum. Group VI—Mercury (ic), Bismuth, Copper, Cadmium. Group VII—Silver, Mercury (ous), Lead. Section II—Acids. Section III—Detection of Bases and Acids. Section IV—Some of the Reactions and Tests of Purity of the more important Organic Compounds. Part III. Quantitative Chemical Analysis. Section I—Gravimetric Estimation. Section II—Volumetric Estimation. There are also a number of useful Tables.

LEFFMANN'S ORGANIC AND MEDICAL CHEMISTRY. Including Urine Analysis and the Analysis of Water and Food. By HENRY LEFFMANN, M.D., Demonstrator of Chemistry at Jefferson Medical College, Philadelphia. 12mo. Cloth, $1.00; Interleaved for the addition of Notes, $1.25

American Health Primers.

EDITED BY W. W. KEEN, M.D.,

Fellow of the College of Physicians of Philadelphia.

This Series of American Health Primers is prepared to diffuse as widely and cheaply as possible, among all classes, a knowledge of the elementary facts of Preventive Medicine, and the bearings and applications of the latest and best researches in every branch of Medical and Hygienic Science. They are intended to teach people the principles of Health, and how to take care of themselves, their children, pupils, employés, etc.

Handsome Cloth Binding, 50 cents, each.

Sent, postpaid, upon receipt of price, or may be obtained from any book store.

HEARING, AND HOW TO KEEP IT. With Illustrations. By CHAS. H. BURNETT, M.D., Aurist to the Presbyterian Hospital, Professor in the Philadelphia Polyclinic.

LONG LIFE, AND HOW TO REACH IT. By J. G. RICHARDSON, M.D., Professor of Hygiene in the University of Pennsylvania.

THE SUMMER AND ITS DISEASES. By JAMES C. WILSON, M.D., Lecturer on Physical Diagnosis in Jefferson Medical College.

EYESIGHT, AND HOW TO CARE FOR IT. With Illustrations. By GEO. C. HARLAN, M.D., Surgeon to the Wills (Eye) Hospital, and to the Eye and Ear Department, Pennsylvania Hospital.

THE THROAT AND THE VOICE. With Illustrations. By J. SOLIS COHEN, M.D., Professor of Diseases of the Throat and Chest in the Philadelphia Polyclinic.

THE WINTER AND ITS DANGERS. By HAMILTON OSGOOD, M.D., of Boston, Editorial Staff Boston *Medical and Surgical Journal.*

THE MOUTH AND THE TEETH. With Illustrations. By J. W. WHITE, M.D., D.D.S., of Philadelphia, Editor of the *Dental Cosmos.*

BRAIN WORK AND OVERWORK. By H. C. WOOD, JR., M.D., Clinical Professor of Nervous Diseases in the University of Pennsylvania.

OUR HOMES. With Illustrations. By HENRY HARTSHORNE, M.D., of Philadelphia, formerly Professor of Hygiene in the University of Pennsylvania.

THE SKIN IN HEALTH AND DISEASE. By L. D. BULKLEY, M.D., of New York, Physician to the Skin Department of the Demilt Dispensary and of the New York Hospital.

SEA AIR AND SEA BATHING. By JOHN H. PACKARD, M.D., of Philadelphia, Surgeon to the Pennsylvania and to St. Joseph's Hospitals.

SCHOOL AND INDUSTRIAL HYGIENE. By D. F. LINCOLN, M.D., of Boston, Chairman Department of Health, American Social Science Association.

"Each volume of the 'American Health Primers' *The Inter-Ocean* has had the pleasure to commend. In their practical teachings, learning, and sound sense, these volumes are worthy of all the compliments they have received. They teach what every man and woman should know, and yet what nine-tenths of the intelligent classes are ignorant of, or at best, have but a smattering knowledge of."—*Chicago Inter-Ocean.*

"The series of American Health Primers deserves hearty commendation. These handbooks of practical suggestion are prepared by men whose professional competence is beyond question, and, for the most part, by those who have made the subject treated the specific study of their lives."

Holden's Manual of Anatomy.

FIFTH EDITION, REVISED AND ENLARGED. 208 ILLUSTRATIONS.

A MANUAL OF THE DISSECTIONS OF THE HUMAN BODY.

By LUTHER HOLDEN, M.D., F.R.C.S., Consulting Surgeon to St. Bartholomew's and the Foundling Hospitals, London, and JOHN LANGTON, F.R.C.S., Surgeon to and Lecturer in St. Bartholomew's Hospital. Fifth Edition. Revised and Enlarged, with many new Illustrations. Octavo. OIL CLOTH BINDING, $4.50 Cloth, $5.00; Leather, $6.00

*** As *Holden's Anatomy* is the chief " Dissector " now in use, the publishers have put it in an Oil Cloth Binding. This does not retain the odors of the dissecting room, is not easily soiled, and may be washed without damage.

DIAGRAM OF AXILLA.

(From Holden's Anatomy.)

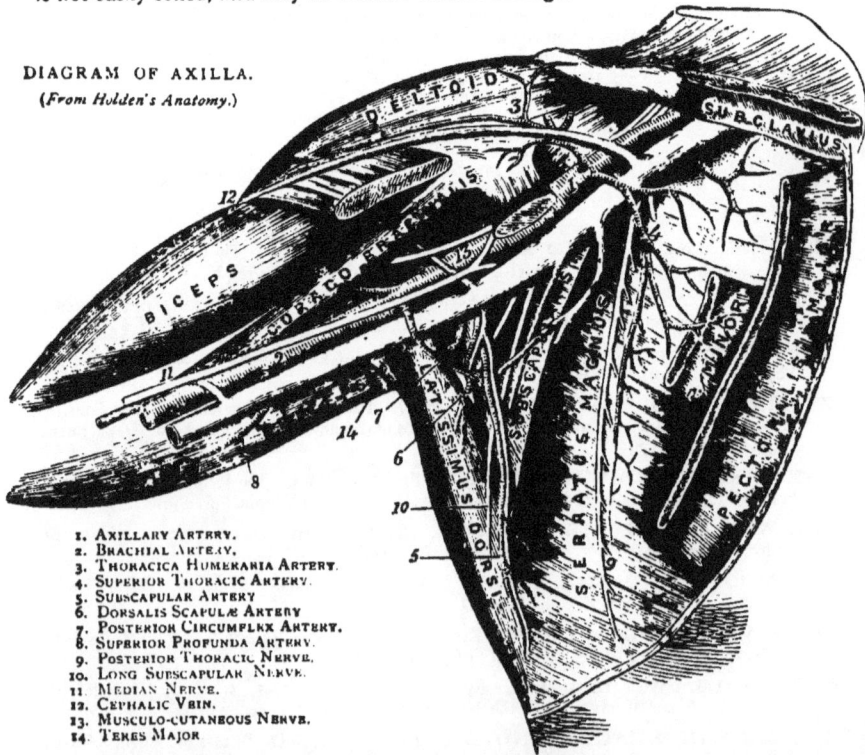

1. AXILLARY ARTERY.
2. BRACHIAL ARTERY.
3. THORACICA HUMERARIA ARTERY.
4. SUPERIOR THORACIC ARTERY.
5. SUBSCAPULAR ARTERY
6. DORSALIS SCAPULÆ ARTERY
7. POSTERIOR CIRCUMFLEX ARTERY.
8. SUPERIOR PROFUNDA ARTERY.
9. POSTERIOR THORACIC NERVE.
10. LONG SUBSCAPULAR NERVE.
11. MEDIAN NERVE.
12. CEPHALIC VEIN.
13. MUSCULO-CUTANEOUS NERVE.
14. TERES MAJOR

" No student of anatomy can take up this book without being pleased and instructed. Its diagrams are original, striking and suggestive, giving more at a glance than pages of text description. All this is known to those who are already acquainted with this admirable work ; but it is simple justice to its value, as a work for careful study and reference, that these points be emphasized to such as are commencing their studies. The text matches the illustrations in directness of practical application and clearness of detail."—*New York Medical Record, April 18th, 1885.*

BY THE SAME AUTHOR.

HUMAN OSTEOLOGY. Comprising a description of the Bones, with Colored Delineations of the Attachments of the Muscles. The General and Microscopical Structure of Bone and its Development. Carefully Revised. By the Author and A. DORAN, F.R.C.S., with Lithographic Plates and Numerous Illustrations. Sixth Edition. 8vo. Cloth, $6.00

HEATH'S PRACTICAL ANATOMY. A Manual of Dissections. Sixth London Edition. 24 Colored Plates, and nearly 300 other Illustrations. Cloth, $5.00

Watson on Amputations.

Amputations of the Extremities and Their Complications. By B. A. WATSON, A.M., M.D., Surgeon to the Jersey City Charity Hospital, to St. Francis' and to Christ's Hospital, at Jersey City, N. J.; Fellow of the American Surgical Association ; Member of the New York Pathological Society, etc. Two full-page Colored Plates, and two hundred and fifty-five Wood Engravings. Octavo. 762 + xix Pages. Handsomely bound in Cloth, $5.50.

SPECIMEN OF ILLUSTRATIONS IN WATSON'S AMPUTATIONS.

"This volume is an encyclopædic monograph, containing the important facts, theories and arguments relating to amputations of the extremities, and their complications. The author's aim has been to collect facts en this subject from English, French, German and American literature. He does not lay claim to originality but, of course, introduces, in their proper places, those observations which his own experience has led him to make on the general subject. Under the division of 'Complications,' the general subject of the treatment of wounds is discussed, with an outline of the present views on germs and germicides, for, in the words of the pref. ce, the complications of amputation wounds are essentially the same as those which pertain to any solution of continuity involving the various tissues of the body. A great service has been done the profession by the insertion of a translation of Gaupot's and Spellmann's writings on artificial limbs. This is the fullest exposition of the subject we have yet seen in an American text-book. Too much praise cannot be given to the typographical appearance of the work. The illustrations are marvels of clearness, and do, what is not always the case, elucidate the text." —*Medical Record, New York, June 20th, 1885.*

Pye's Surgical Handicraft.

A Manual of Surgical Manipulations, Minor Surgery, Bandaging, Dressing, etc., etc., for the use of General Practitioners and Students. With special chapters on Aural Surgery, Extraction of Teeth, Anæsthetics, etc. By WALTER PYE, F.R.C.S., Surgeon to St. Mary's Hospital and the Victoria Hospital for Sick Children ; Examiner in Surgery at the University of Glasgow. 208 Illustrations. Octavo.
Cloth, $5.00.

Heath's Operative Surgery.

A Course of Operative Surgery, consisting of a Series of Colored Plates, each plate containing Several Figures, Drawn from Nature by the Celebrated Anatomical Artist, M. Léveillé, of Paris, Engraved on Steel under his immediate superintendence, with Descriptive Text of Each Operation, and numerous Wood Engravings. By CHRISTOPHER HEATH, F.R.C.S., Surgeon to University College Hospital, and Holme Professor of Clinical Surgery in University College, London. One Large Quarto Volume. Second Edition, Revised and Enlarged. *Sold by Subscription.* *Full information upon application.* Cloth, $12.00.

Practical Handbooks

FOR THE PHYSICIAN AND MEDICAL STUDENT.

VAN HARLINGEN ON SKIN DISEASES. A Handbook of the Diagnosis and Treatment of Skin Diseases. By ARTHUR VAN HARLINGEN, M.D., Professor of Diseases of the Skin in the Philadelphia Polyclinic; Consulting Physician to the Philadelphia Dispensary for Skin Diseases, and Dermatologist to the Howard Hospital. With colored plates representing the appearance of various lesions. 12mo. Cloth, $1.75

*** This is a complete epitome of skin diseases, arranged in alphabetical order, giving the diagnosis and treatment in a concise, practical way. Many prescriptions are given that have never been published in any text-book, and an article incorporated on Diet. The plates do not represent one or two cases, but are composed of a number of figures, accurately colored, showing the appearance of various lesions, and will be found to give great aid in diagnosing.

" This new handbook is essentially a small encyclopædia. * * * Contains a very complete summary of the present state of Dermatology. * * * We heartily commend it for its brevity, clearness and evidently careful preparation."—*Philadelphia Medical Times.*
" The author shows a proper appreciation of the wants of the general practitioner."—*New York Medical Record.*
" It is concisely and intelligently written, and contains many of the best formulas in use for the various forms of Skin Disease."—*New York Medical Times.*
" This is an excellent little book, in which, for ease of reference, the more common diseases of the skin are arranged in alphabetical order, while many good prescriptions are given, together with clear and sensible directions as to their proper application."—*Boston Medical and Surgical Journal.*
" It is just the kind of book that the general practitioner will find most convenient for reference, and we feel confident that it will be appreciated."—*Southern Practitioner.*

RINDFLEISCH'S PATHOLOGY. The Elements of Pathology. By PROF. EDWARD RINDFLEISCH, University of Würzburg. Authorized translation from the first German edition, by WM. H. MERCUR, M.D. (Univ. of Pa.) Revised by JAMES TYSON, M.D , Professor of Pathology and Morbid Anatomy in the University of Pennsylvania. 12mo. Cloth, $2.00

Prof. Tyson, in his Preface to the American edition, says :—" A high appreciation of Prof. Rindfleisch's work on Pathological Histology, caused me to make careful examination of these ' Elements' immediately after their publication in the original. From such an examination I became satisfied that the book would fill a niche in the wants of the student, as well as of others who may desire to familiarize themselves with general pathological processes, viewed from the most modern standpoint."

BRUEN'S PHYSICAL DIAGNOSIS. Second Edition. A Pocket-book of Physical Diagnosis of the Heart and Lungs; for the Student and Physician. By EDWARD T. BRUEN, Demonstrator of Clinical Medicine in the University of Pennsylvania; Lecturer on Pathology in the Women's Medical College of Philadelphia; 2d Edition, revised, with new original illustrations. 12mo. Cloth, $1.50

" We consider the description of the manner and rules governing the art of percussion well given. The subject is always a difficult one for beginners, and requires to be well handled in order to be properly understood."—*American Journal of Medical Sciences.*

WOAKES ON CATARRH AND DISEASES OF THE NOSE CAUSING DEAFNESS. By EDWARD WOAKES, M.D., Senior Aural Surgeon to the London Hospital for Diseases of the Throat and Chest. 29 Illustrations. 12mo. Cloth, $1.50

" Out of the large number of special works on catarrh, there is none for which we have such an unqualified good opinion. * * * The subject is clearly presented. * * * The line of treatment suggested is rational."—*North Carolina Medical Journal.*

VON ARLT. DISEASES OF THE EYE. Including those of the Conjunctiva, Cornea, Sclerotic, and of the Iris and Ciliary Body. By DR. FERDINAND RITTER VON ARLT, Professor of Ophthalmology in Vienna. Translated by LYMAN WARE, M.D., Surgeon to the Illinois Charitable Eye and Ear Infirmary; Ophthalmic Surgeon to the Presbyterian Hospital, and to the Protestant Orphan Asylum, Chicago. Illustrated. 8vo. 325 pages. Cloth, $2.50

"His style is condensed but clear, and his pages contain a vast amount of information, couched in such language that it will be equally instructive to the general practitioner and the specialist."—*Philadelphia Medical and Surgical Reporter, May 30th, 1835.*

TYSON ON THE URINE. A Practical Guide to the Examination of Urine. For the Use of Physicians and Students. With Colored Lithographic Plates and Numerous Illustrations Engraved on Wood. Fourth Edition. 12mo. Cloth, $1.50

"The practical man will find in this little book all that is absolutely necessary for him to know, in order to utilize fully the data supplied by the urine."—*Chicago Medical Journal.*

GILLIAM'S ESSENTIALS OF PATHOLOGY. The Essentials of Pathology. By D. TOD GILLIAM, M.D., Professor of Physiology, Starling Medical College, Columbus, Ohio. With 47 wood engravings. 12mo. Cloth, $2.00

"The general practitioner will find in this little 12mo a convenient compendium of the current pathology of the day."—*Chicago Medical Journal and Examiner.*

THE PRACTICAL SERIES.

A NEW VOLUME JUST READY.

⁎ The volumes of this series written by well known physicians and surgeons, of large private and hospital experience, recognized authorities on the subjects of which they treat, will embrace the various branches of medicine and surgery. They are of a thoroughly practical character, calculated to meet the requirements of the practitioner, and will present the most recent methods and information in a compact shape and at a low price. Bound uniformly, in a handsome and distinctive cloth binding.

DISEASES OF THE KIDNEYS, AND URINARY DERANGEMENTS. By C. H. RALFE, M.A., M.D., F.R.C.P., Assistant Physician to the London Hospital; late Senior Physician to the Seamen's Hospital, Greenwich. 12mo. With Illustrations. 572 pages. *Just Ready.* Cloth, $2.75

BODILY DEFORMITIES AND THEIR TREATMENT. A Handbook of Practical Orthopædics. By H. A. REEVES, F.R.C.S., Senior Assistant Surgeon and Teacher of Practical Surgery at the London Hospital; Surgeon to the Royal Orthopædic Hospital, etc. 12mo. 228 Illustrations. 460 pages. Cloth, $2.25

"From what we have already said, it will be seen that Mr. Reeves has given us a trustworthy guide for the treatment of a very extended class of cases. * * * If the other volumes of the Practical Series are as good as this, we shall be agreeably disappointed."—*American Journal of Medical Sciences, April, 1885.*
"The utility of the work now before us cannot be better recommended to the appreciation of the professional reading public, than by recalling that it is the first of its kind, dealing with orthopædics from a modern standpoint."—*Hospital Gazette and Students' Journal.*

DENTAL SURGERY FOR GENERAL PRACTITIONERS AND STUDENTS IN MEDICINE. By ASHLEY W. BARRETT, M.D., M.R.C.S. ENG., Surgeon-Dentist to, and Lecturer on Dental Surgery and Pathology in the Medical School of, London Hospital. 12mo. Illustrated. Cloth, $1.00

"Replete with an abundance of practical information of unquestionable utility."—*Hospital Gazette and Students' Journal.*
"The object of this volume is to present the student and practitioner with a clear, concise and systematic account of urinary pathology and therapeutics, based upon the latest ascertained facts, and supported by the best authorities. Throughout, the author has endeavored to put prominently forward the characters upon which the diagnosis of the various renal and urinary diseases is founded, and their treatment indicated."—*Extract from The Preface.*

GOODHART AND STARR

ON

The Diseases of Children.

A Manual for Students and Physicians. By J. F. GOODHART, M.D., Physician to the Evelina Hospital for Children; Assistant Physician to Guy's Hospital, London. American Edition, Revised and Edited by LOUIS STARR, M.D., Clinical Professor of Diseases of Children in the Hospital of the University of Pennsylvania, and Physician to the Children's Hospital, Philadelphia. Containing many new Prescriptions, a list of over 50 Formulæ, conforming to the U. S. Pharmacopœia, and Directions for making Artificial Human Milk, for the Artificial Digestion of Milk, etc.

Just Ready. Demi-Octavo. 738 Pages. Cloth, $3.00; Leather, $4.00.

The NEW YORK MEDICAL RECORD, for May, 1885, says:—

" As it is said of some men, so it might be said of some books, that they are ' born to greatness.' This new volume has, we believe, a mission, particularly in the hands of the younger members of the profession. In these days of prolixity in medical literature, it is refreshing to meet with an author who knows both what to say and when he has said it. The work of Dr. Goodhart (admirably conformed, by Dr. Starr, to meet American requirements) is the nearest approach to clinical teaching without the actual presence of clinical material that we have yet seen. It does not discuss mooted questions of Pathology, but is a terse, straightforward account of the author's experience at the bedside of ailing children. Domestic hygiene is awarded its important place in the therapeutics of pediatrics. The details of management so gratefully read by the young practitioner are fully elucidated. Altogether, the book is one of as great practical working value as we have seen for many months."

From the JOURNAL OF THE AMERICAN MEDICAL ASSOCIATION, June 6th, 1885.

" Nothing that concerns disease as found in childhood seems to have escaped the author's attention. From introduction to the end it is replete with valuable information, and one reads it with the feeling that Dr. Goodhart is writing of what he has seen at the bedside. It need scarcely be added that the revisions and additions by the American editor are of much value, neither too full nor too spare, and very judicious."

From the BOSTON MEDICAL AND SURGICAL JOURNAL, June 4th, 1885.

" This work is written in a very agreeable style, carrying weight, from its simplicity and clearness, and the evidently large and matured experience of the author. It is especially adapted to the needs of the practicing physician rather than for the medical student, as with rare discernment it takes up important points in the details of the disease and deals with them practically, leaving the general typical course of the case to the other numerous writers who have already covered the ground in this class of cases. The type and paper are especially to be commended, and the editor, Dr. Starr, can be said to have offered a very attractive book to the medical profession."

From the LONDON MEDICAL TIMES AND GAZETTE, March 7th, 1885.

" Among the great superfluity of medical books which issue from the press we are occasionally gladdened by the reading of some which not only have an unquestionable *raison d' être*, but also as certainly fulfill their purpose. Such a book, we do not hesitate to say, is that which is now before us; and, after a careful perusal, productive of both pleasure and profit, we can assure Dr. Goodhart that he owes no apology for his work, and that if, as he says, he has repeated tales that have been told before, he has repeated them with 'excellent differences.' The book cannot be abstracted. It must, and we think will, be read by all who are interested in or desire to study its subject. We feel sure, moreover, and this forcibly struck us while reading through the work, that those practitioners who have given even half the thought and study to the subject of the diseases of children that Dr. Goodhart has done, will be able to endorse the bulk of his teaching, and will recognize very many of their own unwritten, and sometimes unspoken, thoughts and beliefs; an evidence at once of the great value of the book, and an explanation of the undoubted pleasure that every expert and earnest student will inevitably experience in its perusal."

OTHER WORKS ON DISEASES OF CHILDREN:

DAY. DISEASES OF CHILDREN. A Practical and Systematic Treatise for Practitioners and Students. Second Edition. Rewritten and very much Enlarged. 8vo. 752 pp. Cloth, $3.00; Sheep, $4.00

MEIGS AND PEPPER ON CHILDREN. A Practical Treatise on the Diseases of Children. Seventh Edition, thoroughly Revised and Enlarged.

Cloth, $6.00; Leather, $7.00

? QUIZ-COMPENDS ?

A NEW SERIES OF PRACTICAL MANUALS FOR THE PHYSICIAN AND STUDENT.

Compiled in accordance with the latest teachings of prominent lecturers and the most popular Text-books.

They form a most complete set of Compends, containing information nowhere else collected in such a condensed, practical shape. The authors have had large experience as quiz masters and attachés of colleges, with exceptional opportunities for noting the most recent advances in therapeutics, methods of treatment, etc. The arrangement of the subjects, illustrations and types, are all of the most improved form, and the size of the books is such that they may be easily carried in the pocket.

Bound in Cloth, each $1.00. Interleaved, for the Addition of Notes, $1.25.

No. 1. Human Anatomy. Third Edition. Illustrated. By SAMUEL O. L. POTTER, M.A., M.D., late A. A. Surgeon U. S. Army. With 63 Illus. 3d Revised Ed.

"To those desiring to post themselves hurriedly for examination, this little book will be useful in refreshing the memory."—New Orleans Med. and Surg. Jl.

Nos. 2 and 3. Practice of Medicine. Especially adapted to the use of Students and Physicians. By DANIEL E. HUGHES, M.D., Demonstrator of Clinical Medicine in Jefferson Med. College, Phila. In two parts.

PART I.—Continued, Eruptive and Periodical Fevers, Diseases of the Stomach, Intestines, Peritoneum, Biliary Passages, Liver, Kidneys, etc. (including Tests for Urine), General Diseases, etc.

PART II.—Diseases of the Respiratory System (including Physical Diagnosis), Circulatory System and Nervous System; Diseases of the Blood, etc.

** These little books can be regarded as a full set of notes upon the Practice of Medicine, containing the Synonyms, Definitions, Causes, Symptoms, Prognosis, Diagnosis, Treatment, etc., of each disease, and including a number of prescriptions hitherto unpublished.

No. 4. Physiology, including Embryology. Second Edition. By ALBERT P. BRUBAKER, M.D., Prof. of Physiology, Penn'a College of Dental Surgery; Demonstrator of Physiology in Jefferson Med. College, Phila. Revised and Enlarged.

"This is a well written little book."—London Lancet.

No. 5. Obstetrics. Illustrated. Second Edition. For Physicians and Students. By HENRY G. LANDIS, M.D., Prof. of Obstetrics and Diseases of Women, in Starling Medical College, Columbus. Revised Ed. New Illustrations.

"We have no doubt that many students will find in it a most valuable aid."—The Amer. Jl of Obstetrics.

No. 6. Materia Medica and Therapeutics. Second Revised Edition. With especial Reference to the Physiological Actions of Drugs. For the use of Medical, Dental and Pharmaceutical Students, and Practitioners. Based on the New Revision (Sixth) of the U. S. Pharmacopœia, and including many unofficinal remedies. By SAMUEL O. L. POTTER, M.A., M.D., late A. A. Surg. U. S. Army. Revised Edition, with Index.

"One of the very best we have ever seen."—Southern Clinic.

No. 7. Inorganic Chemistry. New Edition. By G. MASON WARD, M.D., Demonstrator of Chemistry in Jefferson Med. College, Phila. Including Table of Elements and various Analytical Tables. New Ed.

"This neat pocket volume is a brief but excellent compend of inorganic chemistry and simple analysis of the metals."—Pharmaceutical Record, N. Y.

No. 8. Visceral Anatomy. Illustrated. By SAMUEL O. L. POTTER, M.A., M.D., late A. A. Surg. U. S. Army. With 40 Illus.

"Worthy our recommendation to students, and a ready reference to the busy practitioner."—Chicago Med. Times.

No. 9. Surgery. Second Edition. Illustrated. Including Fractures, Wounds, Dislocations, Sprains, Amputations and other operations; Inflammation, Suppuration, Ulcers, Syphilis, Tumors, Shock, etc. Diseases of the Spine, Ear, Eye, Bladder, Testicles, Anus, and other Surgical Diseases. By ORVILLE HORWITZ, A.M., M.D., Resident Physician Pennsylvania Hospital, Phil'a. Second Edition, Revised and Enlarged. With 62 Illustrations.

"Will prove very useful, both to the student and practitioner."—Valentine Mott, M.D., Ass't to the Prof. of Surgery, Bellevue Hospital, New York.

No. 10. Organic Chemistry. Including Medical Chemistry, Urine Analysis, and the Analysis of Water and Food, etc. By HENRY LEFFMANN, M.D., Demonstrator of Chemistry in Jefferson Med. College; Prof. of Chemistry in Penn'a College of Dental Surgery, Philadelphia.

"It is a useful and valuable addition to the series of Quiz-Compends."—College and Clinical Record.

No. 11. Pharmacy. By F. E. STEWART, M.D., PH.G., Quiz Master at Philadelphia College of Pharmacy.

Bound in Cloth, each $1.00. Interleaved, for the Addition of Notes, $1.25.

☞ *These books are constantly revised to keep up with the latest teachings and discoveries.*

www.ingramcontent.com/pod-product-compliance
Lightning Source LLC
Chambersburg PA
CBHW021942190326
41519CB00009B/1111